SpringerBriefs in Applied Sciences and Technology

For further volumes:
http://www.springer.com/series/8884

S. P. Bhattacharyya · L. H. Keel
D. N. Mohsenizadeh

Linear Systems

A Measurement Based Approach

Springer

S. P. Bhattacharyya
Electrical Engineering Department
Texas A&M University
College Station, TX
USA

D. N. Mohsenizadeh
Mechanical Engineering
Texas A&M University
College Station, TX
USA

L. H. Keel
Electrical and Computer Engineering
Tennessee State University
Nashville, TN
USA

ISSN 2191-530X ISSN 2191-5318 (electronic)
ISBN 978-81-322-1640-7 ISBN 978-81-322-1641-4 (eBook)
DOI 10.1007/978-81-322-1641-4
Springer New Delhi Heidelberg New York Dordrecht London

Library of Congress Control Number: 2013947780

© The Author(s) 2014

This work is subject to copyright. All rights are reserved by the Publisher, whether the whole or part of the material is concerned, specifically the rights of translation, reprinting, reuse of illustrations, recitation, broadcasting, reproduction on microfilms or in any other physical way, and transmission or information storage and retrieval, electronic adaptation, computer software, or by similar or dissimilar methodology now known or hereafter developed. Exempted from this legal reservation are brief excerpts in connection with reviews or scholarly analysis or material supplied specifically for the purpose of being entered and executed on a computer system, for exclusive use by the purchaser of the work. Duplication of this publication or parts thereof is permitted only under the provisions of the Copyright Law of the Publisher's location, in its current version, and permission for use must always be obtained from Springer. Permissions for use may be obtained through Rights Link at the Copyright Clearance Center. Violations are liable to prosecution under the respective Copyright Law. The use of general descriptive names, registered names, trademarks, service marks, etc. in this publication does not imply, even in the absence of a specific statement, that such names are exempt from the relevant protective laws and regulations and therefore free for general use. While the advice and information in this book are believed to be true and accurate at the date of publication, neither the authors nor the editors nor the publisher can accept any legal responsibility for any errors or omissions that may be made. The publisher makes no warranty, express or implied, with respect to the material contained herein.

Printed on acid-free paper

Springer is part of Springer Science+Business Media (www.springer.com)

To Gisele, for her love and support

SPB

To My Wife, Kuisook

LHK

To my mother, Shamsozzoha,
my father, Mohammad Farid,
and my brother, Mehrdad

DNM

Preface

This monograph presents the recent results obtained by us on the analysis, synthesis and design of systems described by linear equations. As is well known, linear equations arise in most branches of science and engineering as well as social, biological and economic systems. The novelty of our approach lies in the fact that no models of the system are assumed to be available, nor are they required. Instead, we show that a few measurements made on the system can be processed strategically to directly extract design values that meet specifications without constructing a model of the system, implicitly or explicitly. We illustrate these new concepts by applying them to linear D.C. and A.C. circuits, mechanical, civil and hydraulic systems, signal flow block diagrams and control systems. These applications are preliminary and suggest many open problems. We acknowledge many productive discussions with our colleagues A. Datta, Hazem Nounou, Mohamed Nounou and our graduate students Ritwik Layek and Sirisha Kallakuri.

Earlier research by us has shown that the representation of complex systems by high order models with many parameters may lead to fragility, that is, the drastic change of system behaviour under infinitesimally small perturbations of these parameters. This led to research on model-free measurement-based approaches to design. The results presented in this monograph are our latest effort in this direction and we hope they will lead to attractive alternatives to model-based design of engineering and other systems. We also anticipate applications to robust, adaptive and fault tolerant control.

College Station, USA, June 25, 2013

S. P. Bhattacharyya
L. H. Keel
D. N. Mohsenizadeh

Contents

1 Linear Equations with Parameters 1
  1.1 Introduction 1
  1.2 Parameterized Solutions 3
  1.3 Measurements and Models 4
    1.3.1 Polynomial Models 4
    1.3.2 Rational Models 8
  1.4 Determining a General Parameterized Solution
      from Measurements 10
    1.4.1 A Generalized Superposition Theorem. 11
    1.4.2 A Measurement Theorem. 14
  1.5 Notes and References. 15
  References 16

2 Application to DC Circuits 17
  2.1 Introduction 17
  2.2 Current Control 18
    2.2.1 Current Control Using a Single Resistor 20
    2.2.2 Current Control Using Two Resistors 25
    2.2.3 Current Control Using m Resistors 27
    2.2.4 Current Control Using Gyrator Resistance 29
    2.2.5 Current Control Using m Independent Sources 31
  2.3 Power Level Control 32
    2.3.1 Power Level Control Using a Single Resistor 33
    2.3.2 Power Level Control Using Two Resistors 34
    2.3.3 Power Level Control Using Gyrator Resistance 35
  2.4 Examples of DC Circuit Design 35
  2.5 Notes and References. 42
  References 42

3 Application to AC Circuits 43
  3.1 Current Control 43
    3.1.1 Current Control Using a Single Impedance 44
    3.1.2 Current Control Using Two Impedances 46

       3.1.3   Current Control Using Gyrator Resistance 46

       3.1.4   Current Control Using m Independent Sources 46

  3.2 Power Control. 47

       3.2.1   Power Control Using a Single Impedance 47

       3.2.2   Power Control Using Two Impedances 47

       3.2.3   Power Control Using Gyrator Resistance 48

  3.3 An Example of AC Circuit Design 48

  3.4 Notes and References. 52

4 Application to Mechanical Systems. 53

  4.1 Mass-Spring Systems 53

  4.2 Truss Structures. 55

  4.3 Hydraulic Networks 58

       4.3.1   Flow Rate Control Using a Single Pipe Resistance 59

       4.3.2   Flow Rate Control Using Two Pipe Resistances 61

  4.4 Illustrative Examples 62

       4.4.1   An Example of Mass-Spring Systems 62

       4.4.2   An Example of Truss Structures 64

       4.4.3   An Example of Hydraulic Networks 66

  4.5 Notes and References. 69

  References 69

5 Application to Control Systems 71

  5.1 Introduction 71

  5.2 Block Diagrams. 72

  5.3 SISO Control Systems 73

       5.3.1   Functional Dependency on a Single Controller 73

       5.3.2   Determining a Desired Response 74

       5.3.3   Steps to Controller Design 74

  5.4 An Example of Control System Design 75

  5.5 Notes and References. 80

  References 80

Appendix 81

About the Authors. 85

About the Book. 87

Index 89

Chapter 1
Linear Equations with Parameters

In this chapter, we describe some basic results on the solution of linear equations containing parameters, and the nature of the parameterized solutions. We describe how measurements can be used to extract these parameterized solutions when the equations or models are unknown. These are presented as a generalized Superposition Theorem and a Measurement Theorem.

1.1 Introduction

Consider the system of linear equations

$$\mathbf{Ax} = \mathbf{b}, \tag{1.1}$$

where $\mathbf{A}$ is an $n \times n$ matrix, and $\mathbf{x}$ and $\mathbf{b}$ are $n \times 1$ vectors all with real or complex entries. Let $|.|$ denotes the determinant. Assuming that $|\mathbf{A}| \neq 0$, there exists a unique solution $\mathbf{x}$ and, by Cramer's rule, the ith component x_i of $\mathbf{x}$ is given by

$$x_i = \frac{|\mathbf{A}^i(\mathbf{b})|}{|\mathbf{A}|}, \quad i = 1, 2, \ldots, n \tag{1.2}$$

where $\mathbf{A}^i(\mathbf{b})$ is the matrix obtained by replacing the ith column of $\mathbf{A}$ by $\mathbf{b}$.

In many physical problems, $\mathbf{A}$ and $\mathbf{b}$ contain parameters that need to be chosen or designed, as illustrated in the example below.

Example 1.1. Consider the circuit shown in Fig. 1.1. V is the ideal voltage source, I is the ideal current source, R_1, R_2, R_3 are linear resistors, and R_4 is a gyrator resistance. The gyrator is a linear two port device where the instantaneous currents and the instantaneous voltages are related by $V_2 = R_4 I_2$ and $V_1 = -R_4 I_3$. V_{amp} is the dependent voltage of the amplifier where $V_{amp} = K I_1$, and K is the amplifier gain. The equations of the system can be written in the following matrix form by applying Kirchhoff's current and voltage laws,

S. P. Bhattacharyya et al., *Linear Systems*, SpringerBriefs in Applied Sciences and Technology, DOI: 10.1007/978-81-322-1641-4_1, © The Author(s) 2014

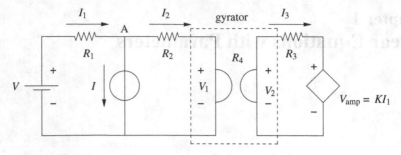

Fig. 1.1 A general circuit

$$
\underbrace{
\begin{bmatrix}
1 & -1 & 0 \\
R_1 & R_2 & -R_4 \\
K & -R_4 & R_3
\end{bmatrix}
}_{\mathbf{A}}
\underbrace{
\begin{bmatrix}
I_1 \\
I_2 \\
I_3
\end{bmatrix}
}_{\mathbf{x}}
=
\underbrace{
\begin{bmatrix}
I \\
V \\
0
\end{bmatrix}
}_{\mathbf{b}}.
\tag{1.3}
$$

To fix notation, we introduce the *parameter* vector $\mathbf{p}$ and the vector of *sources* $\mathbf{q}$:

$$
\mathbf{p} :=
\begin{bmatrix}
R_1 \\
R_2 \\
R_3 \\
R_4 \\
K
\end{bmatrix}
=
\begin{bmatrix}
p_1 \\
p_2 \\
p_3 \\
p_4 \\
p_5
\end{bmatrix}
\quad \text{and} \quad
\mathbf{q} :=
\begin{bmatrix}
I \\
V
\end{bmatrix}
=
\begin{bmatrix}
q_1 \\
q_2
\end{bmatrix},
\tag{1.4}
$$

so that (1.1) can be rewritten showing explicitly the dependence on the parameter vector $\mathbf{p}$ and the source vector $\mathbf{q}$ as

$$
\mathbf{A}(\mathbf{p})\mathbf{x} = \mathbf{b}(\mathbf{q}).
\tag{1.5}
$$

Thus, (1.2) can also be rewritten explicitly showing the parameterized solution as

$$
x_i(\mathbf{p}, \mathbf{q}) = \frac{\left| \mathbf{A}^i(\mathbf{p}, \mathbf{b}(\mathbf{q})) \right|}{|\mathbf{A}(\mathbf{p})|} := \frac{|\mathbf{B}_i(\mathbf{p}, \mathbf{q})|}{|\mathbf{A}(\mathbf{p})|}, \quad i = 1, 2, \ldots, n.
\tag{1.6}
$$

More generally, if

$$
y(\mathbf{p}, \mathbf{q}) = \mathbf{c}^T \mathbf{x}(\mathbf{p}, \mathbf{q}) = c_1 x_1(\mathbf{p}, \mathbf{q}) + \cdots + c_n x_n(\mathbf{p}, \mathbf{q})
\tag{1.7}
$$

is an output of interest, it follows that

$$
y(\mathbf{p}, \mathbf{q}) = \sum_{i=1}^{n} c_i \left(\frac{|\mathbf{B}_i(\mathbf{p}, \mathbf{q})|}{|\mathbf{A}(\mathbf{p})|} \right).
\tag{1.8}
$$

1.2 Parameterized Solutions

Motivated by the above example we consider henceforth the general representation of an arbitrary linear system to be given by (1.5), (1.6) and (1.8). To develop the formula (1.6) in more detail, we note that in (1.3) the parameter $\mathbf{p}$ appears *affinely* in $\mathbf{A}(\mathbf{p})$. Thus, we can write

$$\mathbf{A}(\mathbf{p}) = \mathbf{A}_0 + p_1\mathbf{A}_1 + p_2\mathbf{A}_2 + \cdots + p_l\mathbf{A}_l. \tag{1.9}$$

To proceed, consider the special case of a scalar parameter $\mathbf{p} = p_1$ and

$$\mathbf{A}(\mathbf{p}) = \mathbf{A}_0 + p_1\mathbf{A}_1. \tag{1.10}$$

Lemma 1.1. *With $\mathbf{A}(\mathbf{p})$ as in (1.10), $|\mathbf{A}(\mathbf{p})|$ is a polynomial of degree at most r_1 in p_1 where*

$$r_1 = \text{rank}\,[\mathbf{A}_1]\,. \tag{1.11}$$

Proof. The proof follows easily from the properties of determinants. $\square$

Example 1.2. Consider the following 2×2 matrix $\mathbf{A}$

$$\mathbf{A}(\mathbf{p}) = \begin{bmatrix} 1+p & 1-p \\ p & 2+p \end{bmatrix}, \tag{1.12}$$

which can be written as

$$\mathbf{A}(\mathbf{p}) = \underbrace{\begin{bmatrix} 1 & 1 \\ 0 & 2 \end{bmatrix}}_{\mathbf{A}_0} + p\underbrace{\begin{bmatrix} 1 & -1 \\ 1 & 1 \end{bmatrix}}_{\mathbf{A}_1}. \tag{1.13}$$

Matrix $\mathbf{A}_1$ has rank 2, and we say that matrix $\mathbf{A}(\mathbf{p})$ is of rank 2 with respect to p. Therefore, $|\mathbf{A}(\mathbf{p})|$ will be a polynomial, in p, of degree at most 2, by Lemma 1.1. Calculating $|\mathbf{A}(\mathbf{p})|$ yields

$$|\mathbf{A}(\mathbf{p})| = 2p^2 + 2p + 2. \tag{1.14}$$

Example 1.3. Consider

$$\mathbf{A}(\mathbf{p}) = \begin{bmatrix} 2+p & 1.5+p \\ 1.5+p & 1+p \end{bmatrix} = \underbrace{\begin{bmatrix} 2 & 1.5 \\ 1.5 & 1 \end{bmatrix}}_{\mathbf{A}_0} + p\underbrace{\begin{bmatrix} 1 & 1 \\ 1 & 1 \end{bmatrix}}_{\mathbf{A}_1}, \tag{1.15}$$

so that $\text{rank}[\mathbf{A}_1] = 1$. Here $r_1 = 1$, but

$$|\mathbf{A}(\mathbf{p})| = 0.75, \tag{1.16}$$

is a polynomial of degree 0 in p.

Lemma 1.2. *With* $\mathbf{A}(\mathbf{p})$ *as in* (1.9), *let*

$$r_i = \text{rank}\,[\mathbf{A}_i], \quad i = 1, 2, \ldots, l. \tag{1.17}$$

Then, $|\mathbf{A}(\mathbf{p})|$ *is a multivariate polynomial in* $\mathbf{p}$ *of degree* r_i *or less in* p_i, $i = 1, 2, \ldots, l$
and

$$|\mathbf{A}(\mathbf{p})| = \sum_{i_l=0}^{r_l} \cdots \sum_{i_2=0}^{r_2} \sum_{i_1=0}^{r_1} \alpha_{i_1 i_2 \cdots i_l}\, p_1^{i_1} p_2^{i_2} \cdots p_l^{i_l} := \alpha(\mathbf{p}). \tag{1.18}$$

Also, if the parameter $\mathbf{q}$ *is fixed, say* $\mathbf{q} = \mathbf{q}_0$, *then*

$$|\mathbf{B}_i(\mathbf{p}, \mathbf{q}_0)| = \sum_{i_l=0}^{t_l} \cdots \sum_{i_2=0}^{t_2} \sum_{i_1=0}^{t_1} \beta_{i_1 i_2 \cdots i_l}\, p_1^{i_1} p_2^{i_2} \cdots p_l^{i_l} := \beta_i(\mathbf{p}, \mathbf{q}_0), \tag{1.19}$$

where $\mathbf{B}_i(\mathbf{p}, \mathbf{q}_0)$ *is the matrix obtained by replacing the* ith *column of* $\mathbf{A}(\mathbf{p})$, *in* (1.5),
by the vector $\mathbf{b}(\mathbf{q}_0)$, *and*

$$t_i = \text{rank}\,[\mathbf{B}_i] \le r_i, \quad i = 1, 2, \ldots, l. \tag{1.20}$$

Proof. This follows immediately from Lemma 1.1.   □

Remark 1.1. In the formula (1.18), the number of coefficients $\alpha_{i_1 i_2 \cdots i_l}$ are $\prod_{i=1}^{l}(r_i + 1)$.

Based on the above formula, we have the following characterization of parameterized solutions.

Theorem 1.1. *With* $\mathbf{A}(\mathbf{p})$ *and* $\mathbf{b}(\mathbf{q})$ *as in* (1.5) *and* (1.9),

$$x_i(\mathbf{p}, \mathbf{q}_0) = \frac{\beta_i(\mathbf{p}, \mathbf{q}_0)}{\alpha(\mathbf{p})}, \quad i = 1, 2, \ldots, n, \tag{1.21}$$

where $\beta_i(\mathbf{p}, \mathbf{q}_0)$, $i = 1, 2, \ldots, n$ *and* $\alpha(\mathbf{p})$ *are multivariate polynomials in* $\mathbf{p}$.

Proof. The proof follows from (1.6) and Lemma 1.2.   □

1.3 Measurements and Models

1.3.1 Polynomial Models

In this section we introduce some simple ideas related to measurements and models.
Suppose that the true equation of a system is as follows:

$$y = c_0 + c_1 x + \cdots + c_n x^n, \tag{1.22}$$

where x is an "input" and y is an "output". Now, assume that the mathematical representation (1.22) is not yet known and we are interested in finding a polynomial model of the system by means of measurements. In (1.22) we suppose that the order n and the coefficients $c_i, i = 0, 1, \ldots, n$ are unknown. Consider a polynomial model of the system, of the type:

$$y^{model} = \alpha_0 + \alpha_1 x + \cdots + \alpha_m x^m, \tag{1.23}$$

where x is an "input" and y is an "output" and $\alpha_0, \alpha_1, \ldots, \alpha_m$ are unknown. Suppose experiments are made on the real system by inputting different values $x_1, x_2, \ldots, x_{m+1}$ and measuring the resulting outputs $y_1, y_2, \ldots, y_{m+1}$. The measurement matrix equation is

$$\underbrace{\begin{bmatrix} 1 & x_1 & x_1^2 & \cdots & x_1^m \\ 1 & x_2 & x_2^2 & \cdots & x_2^m \\ \vdots & \vdots & \vdots & & \vdots \\ 1 & x_{m+1} & x_{m+1}^2 & \cdots & x_{m+1}^m \end{bmatrix}}_{\mathbf{M}_m} \underbrace{\begin{bmatrix} \alpha_0 \\ \alpha_1 \\ \vdots \\ \alpha_m \end{bmatrix}}_{\alpha} = \underbrace{\begin{bmatrix} y_1 \\ y_2 \\ \vdots \\ y_{m+1} \end{bmatrix}}_{y}, \tag{1.24}$$

where $|\mathbf{M}_m| = (x_1 - x_2)(x_2 - x_3) \cdots (x_m - x_{m+1})$ and since the input values $x_1, x_2, \ldots, x_{m+1}$ are different, then $|\mathbf{M}_m| \neq 0$. Therefore, (1.24) can be uniquely solved for the vector $\alpha = [\alpha_0, \alpha_1, \ldots, \alpha_m]^T$, and the candidate model (1.23) can be determined. The following three cases are possible:

Case 1: $m = n$.
In this case $\alpha_m \neq 0$, and if another experiment is performed by inputting a different x, called x_{m+2}, and measuring y_{m+2}. Then it must be true that

$$y_{m+2} = \alpha_0 + \alpha_1 x_{m+2} + \cdots + \alpha_m x_{m+2}^m, \tag{1.25}$$

where $\alpha = [\alpha_0, \alpha_1, \ldots, \alpha_m]^T$ is determined from (1.24).

Case 2: $m < n$.
In this situation if another experiment is performed by inputting a different x, called x_{m+2}, then

$$y_{m+2} \neq \alpha_0 + \alpha_1 x_{m+2} + \cdots + \alpha_m x_{m+2}^m. \tag{1.26}$$

In such a case the order of the polynomial model has to be increased.

Case 3: m > n.

In this case $\alpha_m = 0$ (and possibly $\alpha_{m-1} = 0, \ldots$). This fact can be detected from the matrix

$$
\mathbf{M} = \begin{bmatrix} 1 & x_1 & x_1^2 & \cdots & y_1 \\ 1 & x_2 & x_2^2 & \cdots & y_2 \\ \vdots & \vdots & \vdots & & \vdots \\ 1 & x_{m+1} & x_{m+1}^2 & \cdots & y_{m+1} \end{bmatrix},
\tag{1.27}
$$

which must have rank less than $m + 1$. Then one can reduce the order and test the rank of the corresponding matrix again until a full rank matrix is obtained. If one performs another experiment on the system, this model should predict the output correctly; otherwise, we are in Case 2.

Example 1.4. This example illustrates how a polynomial can be constructed (through measurements) to model an actual system. Suppose that the actual system is

$$
y = 2 - x + x^3,
\tag{1.28}
$$

and we begin with the following polynomial model

$$
y^{model} = \alpha_0 + \alpha_1 x,
\tag{1.29}
$$

where α_0 and α_1 are constants to be determined. In order to determine α_0 and α_1, one may perform two experiments by inputting two different values for x into the actual system and measuring the output $y^{measured}$ which is (assumed to be) equal to y. Let us for example set $x_1 = 1$, $x_2 = 2$ and measure $y_1 = 2$, $y_2 = 8$. Thus, we have

$$
\underbrace{\begin{bmatrix} 1 & 1 \\ 1 & 2 \end{bmatrix}}_{\mathbf{M}_2} \underbrace{\begin{bmatrix} \alpha_0 \\ \alpha_1 \end{bmatrix}}_{\alpha} = \underbrace{\begin{bmatrix} 2 \\ 8 \end{bmatrix}}_{y},
\tag{1.30}
$$

which can be solved for $\alpha_0 = -4$ and $\alpha_1 = 6$. Hence,

$$
y^{model} = -4 + 6x.
\tag{1.31}
$$

Suppose we do another experiment using $x_3 = 5$ and measure $y_3 = 122$, but

$$
y^{model}|_{x=5} = 26 \neq 122.
\tag{1.32}
$$

This implies that the order of the polynomial model has to be increased. Now, let us propose

$$
y^{model} = \alpha_0 + \alpha_1 x + \alpha_2 x^2,
\tag{1.33}
$$

and carry out the experiments using $x_1 = 1$, $x_2 = 2$, $x_3 = 5$ which yields $y_1 = 2$, $y_2 = 8$, $y_3 = 122$. Then, we have

$$
\underbrace{\begin{bmatrix} 1 & 1 & 1 \\ 1 & 2 & 4 \\ 1 & 5 & 25 \end{bmatrix}}_{M_3} \underbrace{\begin{bmatrix} \alpha_0 \\ \alpha_1 \\ \alpha_2 \end{bmatrix}}_{\alpha} = \underbrace{\begin{bmatrix} 2 \\ 8 \\ 122 \end{bmatrix}}_{y}, \tag{1.34}
$$

which results in

$$
y^{model} = 12 - 18x + 8x^2. \tag{1.35}
$$

Assume we perform another experiment by setting $x_4 = 7$ and measure $y_4 = 338$, but

$$
y^{model}|_{x=7} = 278 \neq 338. \tag{1.36}
$$

This implies that the order of the polynomial model has to be increased; thus, we consider

$$
y^{model} = \alpha_0 + \alpha_1 x + \alpha_2 x^2 + \alpha_3 x^3. \tag{1.37}
$$

To determine the values of the constants α_0, α_1, α_2 and α_3, let us set $x_1 = 1$, $x_2 = 2$, $x_3 = 5$, $x_4 = 7$ and measure $y_1 = 2$, $y_2 = 8$, $y_3 = 122$, $y_4 = 338$. The following set of measurement equations can be formed:

$$
\underbrace{\begin{bmatrix} 1 & 1 & 1 & 1 \\ 1 & 2 & 4 & 8 \\ 1 & 5 & 25 & 125 \\ 1 & 7 & 49 & 343 \end{bmatrix}}_{M_4} \underbrace{\begin{bmatrix} \alpha_0 \\ \alpha_1 \\ \alpha_2 \\ \alpha_3 \end{bmatrix}}_{\alpha} = \underbrace{\begin{bmatrix} 2 \\ 8 \\ 122 \\ 338 \end{bmatrix}}_{y}, \tag{1.38}
$$

and uniquely solved for the vector of unknown constants $\alpha = [\alpha_0, \alpha_1, \alpha_2, \alpha_3]^T$, which gives $\alpha = [2, -1, 0, 1]^T$ so that

$$
y^{model} = 2 - x + x^3. \tag{1.39}
$$

Let us perform another experiment using $x_5 = 9$ and measure $y_5 = 722$. It can be verified that

$$
y^{model}|_{x=9} = y_5 = 722. \tag{1.40}
$$

Therefore, (1.39) is the correct model.

1.3.2 Rational Models

Here, we extend the idea presented in Sect. 1.3.1 to rational models. Suppose that the mathematical representation of a system is:

$$y = \frac{c_0 + c_1 x + \cdots + c_{m-1} x^{m-1} + c_m x^m}{d_0 + d_1 x + \cdots + d_{n-1} x^{n-1} + x^n}. \tag{1.41}$$

Consider the rational model

$$y^{model} = \frac{\alpha_0 + \alpha_1 x + \cdots + \alpha_s x^s}{\beta_0 + \beta_1 x + \cdots + x^r}. \tag{1.42}$$

Let $k := s + r + 1$, denote the number of unknowns. Similar to the previous case, suppose that experiments are conducted by inputting different values $x_1, x_2, \ldots, x_k$ and measuring the resulting outputs $y_1, y_2, \ldots, y_k$. The measurement matrix equation may be written as

$$\underbrace{\begin{bmatrix} 1 & x_1 & \cdots & x_1^s & -y_1 & -y_1 x_1 & \cdots & -y_1 x_1^{r-1} \\ 1 & x_2 & \cdots & x_2^s & -y_2 & -y_2 x_2 & \cdots & -y_2 x_2^{r-1} \\ \vdots & \vdots & & \vdots & \vdots & \vdots & & \vdots \\ 1 & x_k & \cdots & x_k^s & -y_k & -y_k x_k & \cdots & -y_k x_k^{r-1} \end{bmatrix}}_{\mathbf{M}_k} \underbrace{\begin{bmatrix} \alpha_0 \\ \alpha_1 \\ \vdots \\ \alpha_s \\ \beta_0 \\ \beta_1 \\ \vdots \\ \beta_{r-1} \end{bmatrix}}_{\gamma} = \underbrace{\begin{bmatrix} y_1 x_1^r \\ y_2 x_2^r \\ \vdots \\ y_k x_k^r \end{bmatrix}}_{\mathbf{y}}, \tag{1.43}$$

and can be uniquely solved for the unknowns $\gamma = [\alpha_0, \ldots, \alpha_s, \beta_0, \ldots, \beta_{r-1}]^T$, if and only if $|\mathbf{M}_k| \neq 0$. In a problem of this type, the behavior of the system at large values of the input determines whether the system rational function is proper ($m \leq n$), strictly proper ($m < n$) or improper ($m > n$). Therefore, we consider the following two cases:

Case A: $m \leq n$.
In such a case, if possible, one may conduct at least two experiments for sufficiently large values of the input x. If the output measurements, called y_1 and y_2, are (approximately) equal and $y_1 \approx y_2 \napprox 0$, then $m = n$, and if $y_1 \approx y_2 \approx 0$, then $m < n$. We can write the following

$$y^{model} = y_\infty + y^{strictly\ proper\ model}, \tag{1.44}$$

where y_∞ is the measured value of the system output for a sufficiently large value of the input x, in this case.

Case B: $m > n$.
One can determine if the system rational function is improper by conducting at least two experiments for sufficiently large values of the input x. If the output measurements are not (approximately) equal then $m > n$. In this case the behavior of the system at large values of the input can be modeled as a polynomial function $y^{polynomial\ model}$, using the method presented in Sect. 1.3.1. Hence, we have

$$y^{model} = y^{polynomial\ model} + y^{strictly\ proper\ model}. \tag{1.45}$$

With $y^{polynomial\ model}$ or y_∞ in hand, the problem reduces to determining a strictly proper rational model $y^{strictly\ proper\ model}$. One may consider a model as in (1.42) with $s = r - 1$:

$$y^{strictly\ proper\ model} = \frac{\alpha_0 + \alpha_1 x + \cdots + \alpha_{r-1} x^{r-1}}{\beta_0 + \beta_1 x + \cdots + \beta_r x^r}, \tag{1.46}$$

and determine the coefficients α's and β's by conducting a sufficient number of experiments. As in Sect. 1.3.1, three cases are possible:

Case 1: $r = n$.
In this case $\beta_r \neq 0$, and if another experiment is performed by inputting a different x, say x_{2r+2}, then $y^{model}|_{x_{2r+2}} = y_{2r+2}$.

Case 2: $r < n$.
Here, if one conducts another experience using a different input, for example x_{2r+2}, then $y^{model}|_{x_{2r+2}} \neq y_{2r+2}$. Thus, the order of the numerator and denominator polynomials in (1.46) has to be increased simultaneously by the same value (keeping the relative degree to be 1).

Case 3: $r > n$.
In this situation $\beta_r = 0$ (and possibly $\beta_{r-1} = 0, \ldots$); hence, the order of the denominator and numerator polynomials have to be reduced such that the coefficient of the highest order term in the denominator and numerator are nonzero. If another experiment is done using a different x, called x_{2r+2}, then the predicted value $y^{model}|_{x_{2r+2}}$ is equal to the measured output y_{2r+2}.

Example 1.5. Suppose that the mathematical representation of a system is

$$y = \frac{10}{3 + x^2}. \tag{1.47}$$

Let us, first, examine the behavior of the system at large values of input by inputting $x_1 = 1,000$, $x_2 = 1,200$. We obtain $y_1 \approx y_2 \approx 0$, and thus conclude that $m < n$. Now, consider the following model

$$y^{model} = \frac{\alpha_0}{\beta_0 + x}. \tag{1.48}$$

Let us input two different values for x, for example $x_1 = 1$, $x_2 = 2$, into the system and measure the output y, which yields $y_1 = 2.5$, $y_2 = 1.4286$. Thus, we obtain

$$y = \frac{3.333}{0.333 + x}. \tag{1.49}$$

If we perform another experiment by inputting $x_3 = 4$, we get $y_3 = 0.5263$, but for which $y^{model}|_{x=4} = 0.7692 \neq 0.5263$. Hence, in the next step, we increase the order of the polynomials in the numerator and denominator of (1.48) by 1:

$$y^{model} = \frac{\alpha_0 + \alpha_1 x}{\beta_0 + \beta_1 x + x^2}. \tag{1.50}$$

Table 1.1 summarized the numerical values of the experiments performed for this model.
One can determine the coefficients and obtain

$$y^{model} = \frac{10}{3 + x^2}. \tag{1.51}$$

Now, suppose that one performs another experiment using $x_5 = 10$ and obtains $y_5 = 0.0971$. This can be predicted by substituting $x = 10$ into (1.51) which results in $y^{model}|_{x=10} = y_5 = 0.0971$.

1.4 Determining a General Parameterized Solution from Measurements

In many synthesis and design problems a model is unavailable, but it is necessary to know the behavior of system variable with respect to various parameters. In this

Table 1.1 Numerical values of the experiments performed for example 1.5

Input value	Measured output
1	2.5
2	1.4286
4	0.5263
7	0.1923

section we develop a measurement based approach to the determination of the solution function of a linear system of equations when the model is not known. Specifically, we consider the linear system or model

$$A(p)x = b(q), \qquad (1.52)$$

where the matrix $A(p)$ and vector $b(q)$ are unknown. The objective is to extract the function $x_i(p, q)$ by making measurements of x_i at a set of values of p and q, and processing them strategically to determine the function $x_i(p, q)$. For notational simplicity, drop the suffix i which is fixed and refer to $x_i(p, q)$ as $x(p, q)$. We consider two situations:

(1) Determine $x(q)$ as a function of the sources q, when p is fixed at $p = p_0$,
(2) Determine $x(p)$ as a function of p at a fixed value of $q = q_0$.

1.4.1 A Generalized Superposition Theorem

Here, we suppose that the parameter vector p is fixed at $p = p_0$, and the function $x(q)$ is to be determined from measurements made on the system. To proceed we make the assumption that

$$q = v_1 + v_2 + \cdots + v_k, \qquad (1.53)$$

and $b(q)$ can be decomposed as

$$b(q) = b_1(v_1) + b_2(v_2) + \cdots + b_k(v_k), \qquad (1.54)$$

with

$$b_j(v_j)|_{v_j = 0} = 0, \quad j = 1, 2, \ldots, k. \qquad (1.55)$$

Example 1.6. Suppose $b(q)$ is given as

$$b(q) = \begin{pmatrix} q_1 \\ q_1 q_2 \\ q_3^2 \end{pmatrix}. \qquad (1.56)$$

One can write q as

$$q = \underbrace{\begin{pmatrix} q_1 \\ q_2 \\ 0 \end{pmatrix}}_{v_1} + \underbrace{\begin{pmatrix} 0 \\ 0 \\ q_3 \end{pmatrix}}_{v_2}, \qquad (1.57)$$

giving the following decomposition for $b(q)$

$$\mathbf{b(q)} = \underbrace{\begin{pmatrix} q_1 \\ q_1 q_2 \\ 0 \end{pmatrix}}_{\mathbf{b}_1(\mathbf{v}_1)} + \underbrace{\begin{pmatrix} 0 \\ 0 \\ q_3^2 \end{pmatrix}}_{\mathbf{b}_2(\mathbf{v}_2)},$$ (1.58)

which satisfies (1.55).

The assumptions (1.53), (1.54) and (1.55) allow us to state a general superposition theorem as described below. Let x^j denote the measured value of $x(\mathbf{v}_j)$, $j = 1, 2, \ldots, k$.

Theorem 1.2. (Generalized Superposition Theorem). *Under assumptions* (1.53), (1.54) *and* (1.55)

$$x(\mathbf{q}) = x^1 + x^2 + \cdots + x^k. \tag{1.59}$$

Proof. Let $\mathbf{A}^i(\mathbf{b(q)})$, as before, denote the matrix $\mathbf{A}$ with the ith column replaced by $\mathbf{b(q)}$. Then

$$x(\mathbf{q}) = \frac{|\mathbf{A}^i(\mathbf{b(q)})|}{|\mathbf{A}|}. \tag{1.60}$$

Using (1.53), (1.54) and (1.55), and the fact that

$$|\mathbf{A}^i(\mathbf{b(q)})| = |\mathbf{A}^i(\mathbf{b}_1(\mathbf{v}_1))| + |\mathbf{A}^i(\mathbf{b}_2(\mathbf{v}_2))| + \cdots + |\mathbf{A}^i(\mathbf{b}_k(\mathbf{v}_k))|, \tag{1.61}$$

one can write

$$x(\mathbf{q}) = \frac{|\mathbf{A}^i(\mathbf{b}_1(\mathbf{v}_1))| + |\mathbf{A}^i(\mathbf{b}_2(\mathbf{v}_2))| + \cdots + |\mathbf{A}^i(\mathbf{b}_k(\mathbf{v}_k))|}{|\mathbf{A}|}$$

$$= x^1 + x^2 + \cdots + x^k. \tag{1.62}$$

□

Further simplification results in the special case, in which

$$\mathbf{b(q)} = \mathbf{b}_1 q_1 + \mathbf{b}_2 q_2 + \cdots + \mathbf{b}_m q_m. \tag{1.63}$$

In fact the following theorem is the usual version of the Superposition Theorem one sees in books on circuit theory.

Theorem 1.3. (Superposition Theorem). *Let x^j denote the solution for $x(\mathbf{q})$ when the input is*

$$\mathbf{q}_j := \begin{pmatrix} 0 \\ 0 \\ \vdots \\ q_j \\ \vdots \\ 0 \end{pmatrix}, \quad j = 1, 2, \ldots, m, \tag{1.64}$$

and $x(\mathbf{q})$ the solution for the input

$$\mathbf{q} = \begin{pmatrix} q_1 \\ q_2 \\ \vdots \\ q_m \end{pmatrix} = \mathbf{q}_1 + \mathbf{q}_2 + \cdots + \mathbf{q}_m, \tag{1.65}$$

then

$$x^j = \underbrace{\frac{|\mathbf{A}^i(\mathbf{b}_j)|}{|\mathbf{A}|}}_{\alpha_j} q_j, \tag{1.66}$$

and

$$x(\mathbf{q}) = x^1 + x^2 + \cdots + x^m$$
$$= \alpha_1 q_1 + \alpha_2 q_2 + \cdots + \alpha_m q_m. \tag{1.67}$$

Furthermore, if

$$\mathbf{q} = \begin{pmatrix} c_1 q_1 \\ c_2 q_2 \\ \vdots \\ c_m q_m \end{pmatrix} = c_1 \mathbf{q}_1 + c_2 \mathbf{q}_2 + \cdots + c_m \mathbf{q}_m, \tag{1.68}$$

then

$$x(\mathbf{q}) = c_1 x^1 + c_2 x^2 + \cdots + c_m x^m. \tag{1.69}$$

Proof.   The proof follows from the fact that in the case when (1.68) holds, then

$$|\mathbf{A}^i(\mathbf{b}(\mathbf{q}))| = c_1 |\mathbf{A}^i(\mathbf{b}_1)| q_1 + c_2 |\mathbf{A}^i(\mathbf{b}_2)| q_2 + \cdots + c_m |\mathbf{A}^i(\mathbf{b}_m)| q_m, \tag{1.70}$$

and so

$$x(\mathbf{q}) = c_1 \frac{|\mathbf{A}^i(\mathbf{b}_1)|}{|\mathbf{A}|} q_1 + c_2 \frac{|\mathbf{A}^i(\mathbf{b}_2)|}{|\mathbf{A}|} q_2 + \cdots + c_m \frac{|\mathbf{A}^i(\mathbf{b}_m)|}{|\mathbf{A}|} q_m. \tag{1.71}$$

$\square$

The above theorem states that the solution for the case where the sources are simultaneously active, can be recovered as the sum of the solutions corresponding to the sources acting one at a time.

Remark 1.2. In this case the q_j occur in the matrix $\mathbf{A}^i(\mathbf{b}(\mathbf{q}))$ with rank 1.

Remark 1.3. Note that in Theorem 1.2, if

$$\mathbf{q} = c_1\mathbf{v}_1 + c_2\mathbf{v}_2 + \cdots + c_k\mathbf{v}_k, \tag{1.72}$$

it is in general not true that

$$x(\mathbf{q}) = c_1 x^1 + c_2 x^2 + \cdots + c_k x^k. \tag{1.73}$$

This is because $\mathbf{b}(c_j\mathbf{v}_j) \neq c_j\mathbf{b}(\mathbf{v}_j)$. It is in this sense that the above two theorems are different.

Example 1.7. Recall the circuit example 1.1 where

$$\mathbf{q} = \begin{pmatrix} I \\ V \end{pmatrix} = \underbrace{I}_{q_1} \begin{pmatrix} 1 \\ 0 \end{pmatrix} + \underbrace{V}_{q_2} \begin{pmatrix} 0 \\ 1 \end{pmatrix}. \tag{1.74}$$

The current I_1 can be written as

$$I_1 = \underbrace{\frac{I_1^{1*}}{I^*}}_{\alpha_1} I + \underbrace{\frac{I_1^{2*}}{V^*}}_{\alpha_2} V, \tag{1.75}$$

where α_1 and α_2 can be determined by applying the inputs $[I^*, 0]^T$ and $[0, V^*]^T$, measuring I_1^{1*} and I_1^{2*}, and setting

$$\alpha_1 = \frac{I_1^{1*}}{I^*}, \quad \alpha_2 = \frac{I_1^{2*}}{V^*}. \tag{1.76}$$

1.4.2 A Measurement Theorem

In this section, we consider the determination of the function $x(\mathbf{p}, \mathbf{q}_0) =: x(\mathbf{p})$ when the source vector is fixed at $\mathbf{q} = \mathbf{q}_0$. We exploit the fact that in this case $x(\mathbf{p})$ has the form

$$x(\mathbf{p}) = \frac{\beta(\mathbf{p}, \mathbf{q}_0)}{\alpha(\mathbf{p})} =: \frac{\beta(\mathbf{p})}{\alpha(\mathbf{p})}, \tag{1.77}$$

where $\alpha(\mathbf{p})$ and $\beta(\mathbf{p}, \mathbf{q}_0)$ are defined as in (1.18) and (1.19):

$$\alpha(\mathbf{p}) = |\mathbf{A}(\mathbf{p})| = \sum_{i_l=0}^{r_l} \cdots \sum_{i_2=0}^{r_2} \sum_{i_1=0}^{r_1} \alpha_{i_1 i_2 \cdots i_l} p_1^{i_1} p_2^{i_2} \cdots p_l^{i_l}, \qquad (1.78)$$

$$\beta(\mathbf{p}, \mathbf{q}_0) = |\mathbf{B}(\mathbf{p}, \mathbf{q}_0)| = \sum_{i_l=0}^{t_l} \cdots \sum_{i_2=0}^{t_2} \sum_{i_1=0}^{t_1} \beta_{i_1 i_2 \cdots i_l} p_1^{i_1} p_2^{i_2} \cdots p_l^{i_l}. \qquad (1.79)$$

We, however, assume that for an unknown model ($\mathbf{A}(\mathbf{p})$ and $\mathbf{b}(\mathbf{q})$ not known), the ranks r_i and t_i are known. The coefficients in (1.78) and (1.79), denoted respectively by the vectors α and β, are unknown, and the number of unknown coefficients is $\mu := \prod_{i=1}^{l}(r_i + 1) + \prod_{i=1}^{l}(t_i + 1) - 1$. They can, however, be determined by setting the parameter vector $\mathbf{p}$ to μ different sets of values and solving a set of μ linear equations in the μ unknowns.

Theorem 1.4. (A Measurement Theorem). *The function $x(\mathbf{p})$ can be determined from μ measurements and solution of a system of μ linear equations in the unknown coefficient vectors α and β, called the measurement equations.*

Example 1.8. Suppose that

$$\mathbf{p} = \begin{bmatrix} p_1 \\ p_2 \end{bmatrix}, \quad \mathbf{q} = \begin{bmatrix} q_1 \\ q_2 \end{bmatrix} \quad \text{and} \quad r_1, t_1 = 1, \quad r_2, t_2 = 2.$$

Then,

$$x_i(\mathbf{p}, \mathbf{q}) = \frac{q_1 \beta_{i1}(\mathbf{p}) + q_2 \beta_{i2}(\mathbf{p})}{\alpha(\mathbf{p})}, \qquad (1.80)$$

where

$$\alpha(\mathbf{p}) = \alpha_{00} + \alpha_{10} p_1 + \alpha_{01} p_2 + \alpha_{11} p_1 p_2 + \alpha_{12} p_1 p_2^2, \qquad (1.81)$$

$$\beta_{ij}(\mathbf{p}) = \beta_{ij00} + \beta_{ij10} p_1 + \beta_{ij01} p_2 + \beta_{ij11} p_1 p_2 + \beta_{ij12} p_1 p_2^2, \qquad (1.82)$$

for $i = 1, 2, j = 1, 2$. Now, if p_1, p_2, q_1, q_2 are the design parameters, then the total number of unknown coefficients in (1.80) and thus the number of measurements required to determine the parameterized solution in 5D space of $(p_1, p_2, q_1, q_2, x_i)$ will be 14. In the case where only p_1 and p_2 are the design parameters (q_1 and q_2 are fixed), the number of unknown coefficients is 9 and thus 9 measurements determine the parameterized function. In the situation where p_1 and p_2 are fixed and the design parameters are q_1 and q_2, the number of unknown coefficients becomes 2 and can be determined from two measurements. Note that in (1.80) the dependence of the solution on $\mathbf{p}$ and $\mathbf{q}$ is nonlinear, however, the function can be determined by solving a set of linear equations.

Remark 1.4. This theorem is a generalization of Thevenin's Theorem of circuit theory. This connection will be made in Chaps. 2 and 3. The result given here however applies to any linear system, be it electrical, mechanical, hydraulic and so on.

1.5 Notes and References

In this chapter we developed some formulas expressing the solutions of parameterized sets of linear equations, showing explicitly the nature of the parameter dependence. We also developed a proof of the Superposition Theorem for general linear systems and presented a theorem on the number of measurements required to extract a parameterized solution function for an unknown linear system. These formulas will be useful in subsequent chapters to develop a standardized measurement based approach to the design of linear systems, in various branches of engineering. Some of these applications have been recently explored in [1–6].

References

1. Mohsenizadeh N, Nounou H, Nounou M, Datta A, Bhattacharyya SP (2013) Linear circuits: a measurement based approach. Int J Circuit Theory Appl
2. Layek R, Datta A, Bhattacharyya SP (2011) Linear circuits: a measurement based approach. In: Proceedings of 20th european conference on circuit theory and design, Linkoping, pp 476–479
3. Layek R, Nounou H, Nounou M, Datta A, Bhattacharyya SP (2012) A measurement based approach for linear circuit modeling and design. In: Proceedings of 51th IEEE conference on decision and control, Maui
4. Mohsenizadeh N, Nounou H, Nounou M, Datta A, Bhattacharyya SP (2012) A measurement based approach to circuit design. In: Proceedings of IASTED international conference on engineering and applied science, Colombo, pp 27–34
5. Bhattacharyya SP (2013) Linear sytems: a measurement based approach to analysis, synthesis and design. In: 3rd IASTED Asian conference on modelling, identification and control, Phuket
6. Mohsenizadeh N, Nounou H, Nounou M, Datta A, Bhattacharyya SP (2013) A measurement based approach to mechanical systems. In: Proceedings of 9th Asian control conference, Istanbul

Chapter 2
Application to DC Circuits

In this chapter we use the results obtained in Chap. 1 to develop a new measurement based approach to solve synthesis problems in unknown linear direct current (DC) circuits. We consider two classes of synthesis problems: (1) current control problem, (2) power level control problem. A similar approach can be used for voltage control problems.

2.1 Introduction

Quite often, in a large scale circuit, the detailed model is not available and one may be interested in designing only a small set of, say one, two or three elements, which constitute the design variables. To solve such design problems it is desirable to determine the behavior of the system with respect to these design variables. In this chapter, we provide a new measurement based approach to answer this question by confining ourselves to the domain of linear DC circuits. The approach to be presented can be extended to linear AC circuits, mechanical systems, civil structures, hydraulic networks, transfer functions and parametrized controllers, as we show in later chapters.

Motivated by the question stated above, we pose the problem of determining a circuit signal, such as current, voltage, or power level, in a given branch of an unknown circuit as a function of the design elements located somewhere in the circuit. Note that this functional behavior is nonlinear, in general, even though the underlying circuit is linear.

Of course, the problem of solving the circuit for all the currents can be easily worked out if one knows the circuit model, by applying Kirchhoff's laws to form the linear equations of the system and solving them for the unknown currents. If the circuit model is unavailable, which is often the case in real world design situations, one can resort to experimentally determining this functional dependency by extensive experiments.

S. P. Bhattacharyya et al., *Linear Systems*, SpringerBriefs in Applied Sciences and Technology, DOI: 10.1007/978-81-322-1641-4_2, © The Author(s) 2014

Here, we present an alternative new method which can determine the functional dependency of any circuit variable with respect to any set of design variables directly from a small set of measurements. This has been shown in Chap. 1 to be applicable to any system described by linear equations. The obtained functional dependency can then be used to solve a synthesis problem wherein the circuit variable of interest is to be controlled by adjusting the design variables.

2.2 Current Control

In this section we consider circuit synthesis problems where the current in any branch of an *unknown* linear DC circuit is to be controlled or assigned by adjusting the design elements at arbitrary locations of the circuit. The approach provided here considers several cases where, for example, a single or multiple resistors are used as the design elements. Sources or amplifier gains can also be considered as the design elements.

Let us revisit Example 1.1 (see Fig. 2.1)

In this example, V and I are the ideal voltage and current sources, respectively; R_1, R_2, R_3 are linear branch resistors, and R_4 is a gyrator resistance. In order to make a distinction between a branch resistor and a gyrator resistance, henceforth, we refer to a branch resistor simply as a *resistor* and refer to the latter as a *gyrator resistance*. V_{amp} is the dependent voltage of the amplifier where $V_{\text{amp}} = KI_1$, and K is the amplifier gain. We also introduce the parameter vector and the vector of sources as

$$\mathbf{p} := \begin{bmatrix} R_1 \\ R_2 \\ R_3 \\ R_4 \\ K \end{bmatrix} = \begin{bmatrix} p_1 \\ p_2 \\ p_3 \\ p_4 \\ p_5 \end{bmatrix} \quad \text{and} \quad \mathbf{q} := \begin{bmatrix} I \\ V \end{bmatrix} = \begin{bmatrix} q_1 \\ q_2 \end{bmatrix}. \tag{2.1}$$

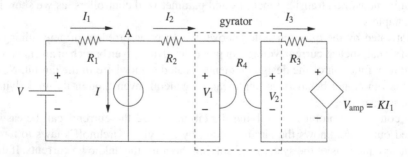

Fig. 2.1 A general circuit

Upon an application of Kirchhoff's current and voltage laws, one can write the equations of the system in the following matrix form,

$$
\underbrace{\begin{bmatrix} 1 & -1 & 0 \\ R_1 & R_2 & -R_4 \\ K & -R_4 & R_3 \end{bmatrix}}_{A(p)} \underbrace{\begin{bmatrix} I_1 \\ I_2 \\ I_3 \end{bmatrix}}_{x} = \underbrace{\begin{bmatrix} I \\ V \\ 0 \end{bmatrix}}_{b(q)}. \tag{2.2}
$$

The governing equations of a linear DC circuit are represented in the following matrix form

$$
A(p)x = b(q), \tag{2.3}
$$

where $A(p)$ is the circuit characteristic matrix, p is the vector of circuit *design parameters*, including resistors, amplifier gains, gyrators, but excluding independent voltage and current sources, x is the vector of unknown currents and q is the vector of independent voltage and current *sources*. The vector $b(q)$ can be written in the following form

$$
b(q) = q_1 b_1 + q_2 b_2 + \cdots + q_m b_m, \tag{2.4}
$$

where $q_1, q_2, \ldots, q_m$ represent the independent sources. Suppose that our objective is to control the current in the i-th branch of the circuit, denoted by I_l. Applying Cramer's rule to (2.3), I_i can be calculated as

$$
x_i = I_i = \frac{|B_i(p, q)|}{|A(p)|}, \tag{2.5}
$$

where $B_i(p, q)$ is the matrix obtained by replacing the i-th column of the characteristic matrix $A(p)$ by the vector $b(q)$. We emphasize that in an unknown circuit the matrices $B_i(p, q)$ and $A(p)$ are unknown. However, based on Lemma 1.2 and (2.4), if the ranks of the parameters are known, a general rational function for the current I_i, in terms of the design elements can be derived, as given below.

$$
I_i = \frac{\sum_{j_m=0}^{1} \cdots \sum_{j_1=0}^{1} \sum_{i_l=0}^{t_l} \cdots \sum_{i_1=0}^{t_1} \alpha_{i_1 \cdots i_l j_1 \cdots j_m} \, p_1^{i_1} \cdots p_l^{i_l} q_1^{j_1} \cdots q_m^{j_m}}{\sum_{i_l=0}^{r_l} \cdots \sum_{i_1=0}^{r_1} \beta_{i_1 \cdots i_l} \, p_1^{i_1} \cdots p_l^{i_l}}. \tag{2.6}
$$

In the above formula, the vectors of numerator and denominator coefficients α and β are constants, $t_1, \ldots, t_l$ are the ranks of the coefficient matrices of the parameters $p_1, \ldots, p_l$ in the matrix $B_i(p, q)$, and $r_1, \ldots, r_l$ are the ranks of the coefficient matrices of the parameters $p_1, \ldots, p_l$ in the matrix $A(p)$.

Fig. 2.2 An unknown linear
DC circuit

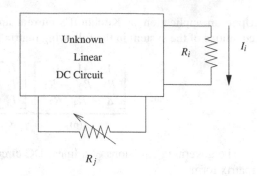

2.2.1 Current Control Using a Single Resistor

Consider the unknown linear DC circuit shown in Fig. 2.2. Suppose that we wish to control the current in the i-th branch, denoted by I_i, by adjusting the resistor R_j at an arbitrary location of the circuit. Following Example 1.1, in general, the resistor R_j will appear in the matrix $\mathbf{A}$ in (2.3) with rank 1 dependency, unless it is a gyrator resistance, in which case the dependency is of rank 2.

Theorem 2.1 *In a linear DC circuit, the functional dependency of any current I_i on any resistance R_j can be determined by at most 3 measurements of the current I_i obtained for 3 different values of R_j.*

Proof. Let us consider two cases: (1) $i \neq j$, and (2) $i = j$.

Case 1: $i \neq j$

In this case, the matrices $\mathbf{B}_i(\mathbf{p}, \mathbf{q})$ and $\mathbf{A}(\mathbf{p})$, in (2.5), are both of rank 1 with respect to R_j. According to Lemma 1.1, the functional dependency of I_i on R_j can be expressed as

$$I_i(R_j) = \frac{\tilde{\alpha}_0 + \tilde{\alpha}_1 R_j}{\tilde{\beta}_0 + \tilde{\beta}_1 R_j}, \tag{2.7}$$

where $\tilde{\alpha}_0, \tilde{\alpha}_1, \tilde{\beta}_0, \tilde{\beta}_1$ are constants. If $\tilde{\beta}_0 = \tilde{\beta}_1 = 0$, then, $I_i \to \infty$, for any value of the resistance R_j, which is physically impossible. Therefore, we rule out this case. Assuming that $\tilde{\beta}_1 \neq 0$, one can divide the numerator and denominator of (2.7) by $\tilde{\beta}_1$ and obtain

$$I_i(R_j) = \frac{\alpha_0 + \alpha_1 R_j}{\beta_0 + R_j}, \tag{2.8}$$

where $\alpha_0, \alpha_1, \beta_0$ are constants. In order to determine $\alpha_0, \alpha_1, \beta_0$ one conducts 3 experiments by setting 3 different values to the resistance R_j, namely R_{j1}, R_{j2}, R_{j3}, and measuring the corresponding currents I_i, namely I_{i1}, I_{i2}, I_{i3}. Then, the following set of measurement equations can be formed

$$\underbrace{\begin{bmatrix} 1 & R_{j1} & -I_{i1} \\ 1 & R_{j2} & -I_{i2} \\ 1 & R_{j3} & -I_{i3} \end{bmatrix}}_{\mathbf{M}} \underbrace{\begin{bmatrix} \alpha_0 \\ \alpha_1 \\ \beta_0 \end{bmatrix}}_{\mathbf{u}} = \underbrace{\begin{bmatrix} I_{i1}R_{j1} \\ I_{i2}R_{j2} \\ I_{i3}R_{j3} \end{bmatrix}}_{\mathbf{m}}. \tag{2.9}$$

The set of Eq. (2.9) can be uniquely solved for the constants $\alpha_0, \alpha_1, \beta_0$, if and only if $|\mathbf{M}| \neq 0$. If $|\mathbf{M}| = 0$, the last column of the matrix $\mathbf{M}$ can be expressed as a linear combination of the first two columns because by assigning different values to the resistance R_j, the first two columns of $\mathbf{M}$ become linearly independent. In such a case, the functional dependency of I_i on R_j can be expressed as

$$I_i(R_j) = \alpha_0 + \alpha_1 R_j, \tag{2.10}$$

where α_0, α_1 are constants that can be determined from any two of the experiments conducted earlier. The functional dependency in (2.10) corresponds to the case where $\tilde{\beta}_1 = 0$ in (2.7), and the numerator and denominator of (2.7) are divided by $\tilde{\beta}_0$.

Case 2: $i = j$

In this case, the matrix $\mathbf{A}(\mathbf{p})$ is of rank 1 with respect to R_i; however, the matrix $\mathbf{B}_i(\mathbf{p}, \mathbf{q})$ is of rank 0 with respect to R_i. According to Lemma 1.1, the functional dependency of I_i on R_i can be expressed as

$$I_i(R_i) = \frac{\tilde{\alpha}_0}{\tilde{\beta}_0 + \tilde{\beta}_1 R_i}, \tag{2.11}$$

where $\tilde{\alpha}_0, \tilde{\beta}_0, \tilde{\beta}_1$ are constants. Assuming that $\tilde{\beta}_1 \neq 0$, and dividing the numerator and denominator of (2.11) by $\tilde{\beta}_1$, gives

$$I_i(R_i) = \frac{\alpha_0}{\beta_0 + R_i}, \tag{2.12}$$

where α_0, β_0 are constants that can be determined by conducting 2 experiments, by setting 2 different values to the resistance R_i, namely R_{i1}, R_{i2}, and measuring the corresponding currents I_i, namely I_{i1}, I_{i2}. The following set of measurement equations can then be formed

$$\underbrace{\begin{bmatrix} 1 & -I_{i1} \\ 1 & -I_{i2} \end{bmatrix}}_{\mathbf{M}} \underbrace{\begin{bmatrix} \alpha_0 \\ \beta_0 \end{bmatrix}}_{\mathbf{u}} = \underbrace{\begin{bmatrix} I_{i1}R_{i1} \\ I_{i2}R_{i2} \end{bmatrix}}_{\mathbf{m}}. \tag{2.13}$$

The set of Eq. (2.13) can be uniquely solved for the constants α_0, β_0, provided $|\mathbf{M}| \neq 0$. If $|\mathbf{M}| = 0$ in (2.13), it can be concluded that I_i is a constant,

$$I_i(R_i) = \alpha_0, \tag{2.14}$$

Fig. 2.3 Graph of (2.8) for $\beta_0 > 0$, $\alpha_0 < 0$ and $\alpha_1 > 0$

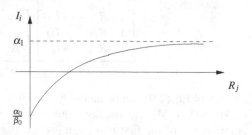

which can be determined from any of the experiments conducted earlier. In this case, the functional dependency in (2.14) corresponds to the situation where $\tilde{\beta}_1 = 0$ in (2.11), and the numerator and denominator of (2.11) are divided by $\tilde{\beta}_0$. $\square$

Remark 2.1 Suppose that $i \neq j$ and $|\mathbf{M}| \neq 0$ in (2.9), then the derivative of $I_i(R_j)$ in (2.8), with respect to R_j, can be calculated as

$$\frac{dI_i}{dR_j} = \frac{\alpha_1 \beta_0 - \alpha_0}{(\beta_0 + R_j)^2}. \tag{2.15}$$

If $\beta_0 \geq 0$, we have the following:

1. The function (2.8) is monotonic in R_j, i.e. $I_i(R_j)$ monotonically increases or decreases as R_j increases from 0 to large values. The limiting values of this function are: $I_i(0) = \frac{\alpha_0}{\beta_0}$ and $I_i(\infty) = \alpha_1$. If $\frac{\alpha_0}{\beta_0} > \alpha_1$, then (2.8) will monotonically decrease, and if $\frac{\alpha_0}{\beta_0} < \alpha_1$, then (2.8) will monotonically increase.
2. The achievable range for I_i, by varying R_j in the interval $[0, \infty)$, is

$$\min \left\{ \frac{\alpha_0}{\beta_0}, \alpha_1 \right\} < I_i < \max \left\{ \frac{\alpha_0}{\beta_0}, \alpha_1 \right\}. \tag{2.16}$$

3. In a current control problem of this type, this monotonic behavior allows us to uniquely determine a range of values of the design parameter R_j, $R_j^- \leq R_j \leq R_j^+$, for which the current I_i lies within a desired prescribed range, $I_i^- \leq I_i \leq I_i^+$, which of course must be within the achievable range (2.16).

These observations also are clear from the graph of (2.8). For instance, if $\beta_0 > 0$, $\alpha_0 < 0$ and $\alpha_1 > 0$, the graph of (2.8) has the general shape as depicted in Fig. 2.3.

If $\beta_0 < 0$, we have:

1. The function (2.8) is monotonic in R_j, in the intervals $[0, -\beta_0)$ and $(-\beta_0, \infty)$. If $\alpha_1 \beta_0 - \alpha_0 > 0$, then I_i starts at $\frac{\alpha_0}{\beta_0}$ and monotonically increases to $+\infty$ as $R_j \to -\beta_0^-$, then, at $R_j \to -\beta_0^+$ it starts from $-\infty$ and monotonically increases to α_1 as $R_j \to \infty$. If $\alpha_1 \beta_0 - \alpha_0 < 0$, I_i starts at $\frac{\alpha_0}{\beta_0}$ and monotonically decreases to $-\infty$ as $R_j \to -\beta_0^-$, then, at $R_j \to -\beta_0^+$ it starts from $+\infty$ and monotonically decreases to α_1 as $R_j \to \infty$.

2. The achievable range for I_i, by varying R_j in the interval $[0, \infty)$, is

$$I_i \in \left(-\infty, \min\left\{\frac{\alpha_0}{\beta_0}, \alpha_1\right\}\right) \cup \left(\max\left\{\frac{\alpha_0}{\beta_0}, \alpha_1\right\}, +\infty\right). \qquad (2.17)$$

3. Similarly, one can uniquely determine a range of values of the design parameter R_j, $R_j^- \le R_j \le R_j^+$, for which the current I_i lies within a desired prescribed range, $I_i^- \le I_i \le I_i^+$, which of course must be within the achievable range (2.17).

The graph of (2.8) for this case again clearly illustrates (see Fig. 2.4).

Thevenin's Theorem (the special case $i = j$)
Thevenin's Theorem of circuit theory follows as a special case of the results developed here. To see this, consider the current functional dependency given in (2.12). From this relationship, it is clear that the short circuit current I_{sc} is given by $I_{sc} = \frac{\alpha_0}{\beta_0}$, which is obtained by setting $R_i = 0$. Similarly, the open circuit voltage V_{oc} is obtained by multiplying both sides of (2.12) by R_i and taking the limit as $R_i \to \infty$. This yields $V_{oc} = V_{Th} = \alpha_0$. Thus, the Thevenin resistance is given by $R_{Th} = \frac{V_{oc}}{I_{sc}} = \beta_0$, so that (2.12) becomes

$$I_i(R_i) = \frac{V_{Th}}{R_{Th} + R_i}, \qquad (2.18)$$

which is exactly Thevenin's Theorem. We point out that in our approach, it is not necessary to measure short circuit current or open circuit voltage; indeed two *arbitrary* measurements suffice. This has practical and useful implications in circuits where short circuiting and open circuiting may sometimes be impossible.

Remark 2.2 (Generalization of Thevenin's Theorem) Theorem 2.1 and the subsequent results in this chapter represent generalizations of Thevenin's Theorem. In Thevenin's Theorem, the current in a resistor/impedance connected to an arbitrary network can be obtained by representing the network by a voltage source and a resistance/impedance and these can be determined from short circuit and open circuit measurements made at these terminals. We have shown that the resistor can be connected at a point different from the point where measurements are made and that the current can be predicted from arbitrary measurements, not necessarily short

Fig. 2.4 Graph of (2.8) for $\beta_0 < 0$, $\alpha_0 > 0$ and $\alpha_1 > 0$

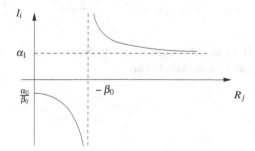

or open circuit. The results given in the subsequent chapters can be thought of as Thevenin-like results for mechanical, hydraulic, truss and other systems.

Current Assignment Problem

Once the functional dependency of interest, (2.8), (2.10), (2.12) or (2.14) is obtained, a synthesis problem can be solved. For instance, suppose that it is required to assign $I_i = I_i^*$, where I_i^* is a desired prescribed value of the current in the i-th branch of the unknown circuit. Let us assume that the design variable is the resistance R_j, and $i \neq j$. How can one find a value of R_j for which $I_i = I_i^*$? Based on the Theorem 2.1, since $i \neq j$, one conducts 3 experiments by setting 3 different values to the resistance R_j, and measuring the corresponding currents I_i. The matrix $\mathbf{M}$ in (2.9) can then be evaluated from the measurements. If $|\mathbf{M}| \neq 0$, then the functional dependency of interest will be of the form given in (2.8), and if $|\mathbf{M}| = 0$, it will be of the form obtained in (2.10). Suppose that $|\mathbf{M}| \neq 0$ is the case; hence, the functional dependency of interest is as the one given in (2.8). In order to determine the value of R_j, for which the desired current I_i^* is attained, one may solve (2.8) for R_j, with $I_i = I_i^*$,

$$R_j(I_i^*) = \frac{\alpha_0 - I_i^* \beta_0}{I_i^* - \alpha_1}. \tag{2.19}$$

Interval Design Problem

Suppose now that the current I_i is to be controlled to stay within the following range (which is inside the achievable range (2.16)),

$$I_i^- \leq I_i \leq I_i^+, \tag{2.20}$$

by adjusting the design resistance R_j, $i \neq j$. Also, assume that after conducting 3 experiments, we found $|\mathbf{M}| \neq 0$ in (2.9) and $\beta_0 \geq 0$. Therefore, the functional dependency of I_i on R_j is of the form (2.8) and is monotonic. Thus, one may find a unique corresponding interval for R_j values where (2.20) is satisfied. Supposing that I_i, in (2.8), monotonically increases as R_j increases, one gets

$$R_j^- \leq R_j \leq R_j^+, \tag{2.21}$$

where

$$R_j^- = \frac{\alpha_0 - I_i^- \beta_0}{I_i^- - \alpha_1}, \qquad R_j^+ = \frac{\alpha_0 - I_i^+ \beta_0}{I_i^+ - \alpha_1}. \tag{2.22}$$

If I_i, in (2.8), monotonically decreases as R_j increases; then, R_j^- and R_j^+ in (2.21) can be calculated from

$$R_j^- = \frac{\alpha_0 - I_i^+ \beta_0}{I_i^+ - \alpha_1}, \qquad R_j^+ = \frac{\alpha_0 - I_i^- \beta_0}{I_i^- - \alpha_1}. \tag{2.23}$$

Following the same strategy, one may solve a synthesis problem for the case $i = j$. The problem of maintaining several currents in the circuit within prescribed intervals can be solved similarly. Also, the discussion above pertains to detection of short and open circuit faults.

2.2.2 Current Control Using Two Resistors

Consider the unknown linear DC circuit shown in Fig. 2.5. Suppose we want to control the current in the i-th branch, denoted by I_i, by adjusting any two resistors R_j and R_k at arbitrary locations of the circuit. Assume, as before, that R_j and R_k are not gyrator resistances.

Theorem 2.2 *In a linear DC circuit, the functional dependency of any current I_i on any two resistances R_j and R_k can be determined by at most 7 measurements of the current I_i obtained for 7 different sets of values (R_j, R_k).*

Proof. Let us consider two cases: (1) $i \neq j, k$ and (2) $i = j$ or $i = k$.

Case 1: $i \neq j, k$

In this case, the matrices $\mathbf{B}_i(\mathbf{p}, \mathbf{q})$ and $\mathbf{A}(\mathbf{p})$, in (2.5), are both of rank 1 with respect to R_j and R_k. Based on Lemma 1.2, the functional dependency of I_i on R_j and R_k can be expressed as

$$I_i(R_j, R_k) = \frac{\tilde{\alpha}_0 + \tilde{\alpha}_1 R_j + \tilde{\alpha}_2 R_k + \tilde{\alpha}_3 R_j R_k}{\tilde{\beta}_0 + \tilde{\beta}_1 R_j + \tilde{\beta}_2 R_k + \tilde{\beta}_3 R_j R_k}, \qquad (2.24)$$

where $\tilde{\alpha}_0, \tilde{\alpha}_1, \tilde{\alpha}_2, \tilde{\alpha}_3, \tilde{\beta}_0, \tilde{\beta}_1, \tilde{\beta}_2, \tilde{\beta}_3$ are constants. Assuming that $\tilde{\beta}_3 \neq 0$ and dividing the numerator and denominator of (2.24) by $\tilde{\beta}_3$, yields

$$I_i(R_j, R_k) = \frac{\alpha_0 + \alpha_1 R_j + \alpha_2 R_k + \alpha_3 R_j R_k}{\beta_0 + \beta_1 R_j + \beta_2 R_k + R_j R_k}, \qquad (2.25)$$

Fig. 2.5 An unknown linear DC circuit

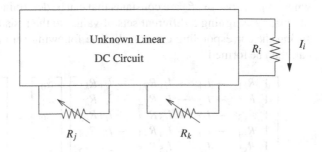

where $\alpha_0, \alpha_1, \alpha_2, \alpha_3, \beta_0, \beta_1, \beta_2$ are constants. In order to determine these constants, one conducts 7 experiments by assigning 7 different sets of values to the resistances (R_j, R_k), and measuring the corresponding currents I_i. The following set of measurement equations will be obtained

$$
\underbrace{\begin{bmatrix}
1 & R_{j1} & R_{k1} & R_{j1}R_{k1} & -I_{i1} & -I_{i1}R_{j1} & -I_{i1}R_{k1} \\
1 & R_{j2} & R_{k2} & R_{j2}R_{k2} & -I_{i2} & -I_{i2}R_{j2} & -I_{i2}R_{k2} \\
1 & R_{j3} & R_{k3} & R_{j3}R_{k3} & -I_{i3} & -I_{i3}R_{j3} & -I_{i3}R_{k3} \\
1 & R_{j4} & R_{k4} & R_{j4}R_{k4} & -I_{i4} & -I_{i4}R_{j4} & -I_{i4}R_{k4} \\
1 & R_{j5} & R_{k5} & R_{j5}R_{k5} & -I_{i5} & -I_{i5}R_{j5} & -I_{i5}R_{k5} \\
1 & R_{j6} & R_{k6} & R_{j6}R_{k6} & -I_{i6} & -I_{i6}R_{j6} & -I_{i6}R_{k6} \\
1 & R_{j7} & R_{k7} & R_{j7}R_{k7} & -I_{i7} & -I_{i7}R_{j7} & -I_{i7}R_{k7}
\end{bmatrix}}_{\mathbf{M}}
\underbrace{\begin{bmatrix}
\alpha_0 \\ \alpha_1 \\ \alpha_2 \\ \alpha_3 \\ \beta_0 \\ \beta_1 \\ \beta_2
\end{bmatrix}}_{\mathbf{u}}
=
\underbrace{\begin{bmatrix}
I_{i1}R_{j1}R_{k1} \\
I_{i2}R_{j2}R_{k2} \\
I_{i3}R_{j3}R_{k3} \\
I_{i4}R_{j4}R_{k4} \\
I_{i5}R_{j5}R_{k5} \\
I_{i6}R_{j6}R_{k6} \\
I_{i7}R_{j7}R_{k7}
\end{bmatrix}}_{\mathbf{m}}. \quad (2.26)
$$

This set of equations can be uniquely solved for the constants $\alpha_0, \alpha_1, \alpha_2, \alpha_3$, $\beta_0, \beta_1, \beta_2$, if and only if $|\mathbf{M}| \neq 0$ in (2.26). In the case where $|\mathbf{M}| = 0$, one can follow the same procedure used in Sect. 2.2.1 to derive the corresponding functional dependency of I_i on R_j and R_k. We provide the details of this case in the Appendix.

Case 2: $i = j$ or $i = k$

Suppose that $i = j$ and recall (2.5). In this case, the matrix $\mathbf{A}(\mathbf{p})$ is of rank 1 with respect to R_i and R_k; however, the matrix $\mathbf{B}_i(\mathbf{p}, \mathbf{q})$ is of rank 0 with respect to R_i and is of rank 1 with respect to R_k. According to Lemma 1.2 and these rank conditions, the functional dependency of I_i on R_i and R_k can be expressed as

$$
I_i(R_i, R_k) = \frac{\tilde{\alpha}_0 + \tilde{\alpha}_1 R_k}{\tilde{\beta}_0 + \tilde{\beta}_1 R_i + \tilde{\beta}_2 R_k + \tilde{\beta}_3 R_i R_k}, \quad (2.27)
$$

where $\tilde{\alpha}_0, \tilde{\alpha}_1, \tilde{\beta}_0, \tilde{\beta}_1, \tilde{\beta}_2, \tilde{\beta}_3$ are constants. Assuming that $\tilde{\beta}_3 \neq 0$, one can divide the numerator and denominator of (2.27) by $\tilde{\beta}_3$ and obtain

$$
I_i(R_i, R_k) = \frac{\alpha_0 + \alpha_1 R_k}{\beta_0 + \beta_1 R_i + \beta_2 R_k + R_i R_k}, \quad (2.28)
$$

where $\alpha_0, \alpha_1, \beta_0, \beta_1, \beta_2$ are constants that can be determined by conducting 5 experiments, by assigning 5 different sets of values to the resistances (R_i, R_k), and measuring the corresponding currents I_i. The following set of measurement equations can then be formed

$$
\underbrace{\begin{bmatrix}
1 & R_{k1} & -I_{i1} & -I_{i1}R_{j1} & -I_{i1}R_{k1} \\
1 & R_{k2} & -I_{i2} & -I_{i2}R_{j2} & -I_{i2}R_{k2} \\
1 & R_{k3} & -I_{i3} & -I_{i3}R_{j3} & -I_{i3}R_{k3} \\
1 & R_{k4} & -I_{i4} & -I_{i4}R_{j4} & -I_{i4}R_{k4} \\
1 & R_{k5} & -I_{i5} & -I_{i5}R_{j5} & -I_{i5}R_{k5}
\end{bmatrix}}_{\mathbf{M}}
\underbrace{\begin{bmatrix}
\alpha_0 \\ \alpha_1 \\ \beta_0 \\ \beta_1 \\ \beta_2
\end{bmatrix}}_{\mathbf{u}}
=
\underbrace{\begin{bmatrix}
I_{i1}R_{j1}R_{k1} \\
I_{i2}R_{j2}R_{k2} \\
I_{i3}R_{j3}R_{k3} \\
I_{i4}R_{j4}R_{k4} \\
I_{i5}R_{j5}R_{k5}
\end{bmatrix}}_{\mathbf{m}}. \quad (2.29)
$$

Again, this set of equations can be uniquely solved for the constants α_0, α_1, $\beta_0, \beta_1, \beta_2$, provided $|\mathbf{M}| \neq 0$ in (2.29). If $|\mathbf{M}| = 0$, following the same strategy used in Sect. 2.2.1, one can derive the corresponding functional dependency of I_i on R_i and R_k. The details of this case can be found in the Appendix. □

In this problem, the current I_i can be plotted as a surface in a 3D graph. In a synthesis problem of this type, any constraint on the current I_i results in a corresponding region in the R_j–R_k plane, if the solution set for that constraint is not empty.

2.2.3 Current Control Using m Resistors

Consider the unknown linear DC circuit shown in Fig. 2.6.
Suppose that the objective is to control the current in the i-th branch of the circuit, denoted by I_i, by adjusting any m resistors R_j, $j = 1, 2, \ldots, m$, at arbitrary locations of the circuit. Assume that the resistances R_j, $j = 1, 2, \ldots, m$, are not gyrator resistances.

Theorem 2.3 *In a linear DC circuit, the functional dependency of any current I_i on any m resistances R_j, $j = 1, 2, \ldots, m$, can be determined by at most $2^{m+1} - 1$ measurements of the current I_i obtained for $2^{m+1} - 1$ different sets on values of the vector $(R_1, R_2, \ldots, R_m)$.*

Proof. Let us consider two cases: (1) $i \neq j$ for $j = 1, 2, \ldots, m$, and (2) $i = j$ for some $j = 1, 2, \ldots, m$.

Case 1: $i \neq j$, $j = 1, 2, \ldots, m$

In this case, the matrices $\mathbf{B}_i(\mathbf{p}, \mathbf{q})$ and $\mathbf{A}(\mathbf{p})$, in (2.5), are both of rank 1 with respect to R_j, $j = 1, 2, \ldots, m$. Hence, based on Lemma 1.2, the functional dependency of I_i on $R_1, R_2, \ldots, R_m$ can be written as

$$I_i(R_1, R_2, \ldots, R_m) = \frac{\sum_{i_m=0}^{1} \cdots \sum_{i_2=0}^{1} \sum_{i_1=0}^{1} \tilde{\alpha}_{i_1 i_2 \cdots i_m} \, R_1^{i_1} R_2^{i_2} \cdots R_m^{i_m}}{\sum_{i_m=0}^{1} \cdots \sum_{i_2=0}^{1} \sum_{i_1=0}^{1} \tilde{\beta}_{i_1 i_2 \cdots i_m} \, R_1^{i_1} R_2^{i_2} \cdots R_m^{i_m}}, \quad (2.30)$$

Fig. 2.6 An unknown linear DC circuit

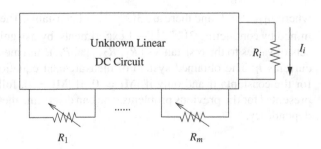

where $\tilde{\alpha}_{i_1 i_2 \cdots i_m}$'s and $\tilde{\beta}_{i_1 i_2 \cdots i_m}$'s are constants. Assuming that $\tilde{\beta}_{11 \cdots 1} \neq 0$ and dividing the numerator and denominator of (2.30) by $\tilde{\beta}_{11 \cdots 1}$, results in

$$I_i(R_1, R_2, \ldots, R_m) = \frac{\sum_{i_m=0}^{1} \cdots \sum_{i_2=0}^{1} \sum_{i_1=0}^{1} \alpha_{i_1 i_2 \cdots i_m} R_1^{i_1} R_2^{i_2} \cdots R_m^{i_m}}{\sum_{i_m=0}^{1} \cdots \sum_{i_2=0}^{1} \sum_{i_1=0}^{1} \beta_{i_1 i_2 \cdots i_m} R_1^{i_1} R_2^{i_2} \cdots R_m^{i_m}}, \quad (2.31)$$

where $\beta_{11 \cdots 1} = 1$, and $\alpha_{i_1 i_2 \cdots i_m}$'s and $\beta_{i_1 i_2 \cdots i_m}$'s are $2^{m+1} - 1$ constants. In order to determine these constants, one conducts $2^{m+1} - 1$ experiments, by setting $2^{m+1} - 1$ different sets of values to the resistances $(R_1, R_2, \ldots, R_m)$, and measuring the corresponding currents I_i. The obtained set of measurement equations has a unique solution for the constants if and only if $|\mathbf{M}| \neq 0$. If $|\mathbf{M}| = 0$ is the case, then one can follow the same procedure explained for the previous problems to derive the corresponding functional dependency.

Case 2: $i = j$ for some $j = 1, 2, \ldots, m$

Without loss of generality, suppose that $i = m$ and recall (2.5). In this case, the matrix $\mathbf{A}(\mathbf{p})$ is of rank 1 with respect to R_j, $j = 1, 2, \ldots, m$; however, the matrix $\mathbf{B}_i(\mathbf{p}, \mathbf{q})$ is of rank 0 with respect to R_m and is of rank 1 with respect to R_j, $j = 1, 2, \ldots, m - 1$. According to these rank conditions and based on Lemma 1.2, the functional dependency of I_i on $R_1, R_2, \ldots, R_m$ will be

$$I_i(R_1, R_2, \ldots, R_m) = \frac{\sum_{i_{m-1}=0}^{1} \cdots \sum_{i_2=0}^{1} \sum_{i_1=0}^{1} \tilde{\alpha}_{i_1 i_2 \cdots i_{m-1}} R_1^{i_1} R_2^{i_2} \cdots R_{m-1}^{i_{m-1}}}{\sum_{i_m=0}^{1} \cdots \sum_{i_2=0}^{1} \sum_{i_1=0}^{1} \tilde{\beta}_{i_1 i_2 \cdots i_m} R_1^{i_1} R_2^{i_2} \cdots R_m^{i_m}},$$

$$(2.32)$$

where $\tilde{\alpha}_{i_1 i_2 \cdots i_{m-1}}$'s and $\tilde{\beta}_{i_1 i_2 \cdots i_m}$'s are constants. Supposing $\tilde{\beta}_{11 \cdots 1} \neq 0$, one can divide the numerator and denominator of (2.32) by $\tilde{\beta}_{11 \cdots 1}$ and get

$$I_i(R_1, R_2, \ldots, R_m) = \frac{\sum_{i_{m-1}=0}^{1} \cdots \sum_{i_2=0}^{1} \sum_{i_1=0}^{1} \alpha_{i_1 i_2 \cdots i_{m-1}} R_1^{i_1} R_2^{i_2} \cdots R_{m-1}^{i_{m-1}}}{\sum_{i_m=0}^{1} \cdots \sum_{i_2=0}^{1} \sum_{i_1=0}^{1} \beta_{i_1 i_2 \cdots i_m} R_1^{i_1} R_2^{i_2} \cdots R_m^{i_m}},$$

$$(2.33)$$

where $\beta_{11 \cdots 1} = 1$, and there are $3(2^{m-1}) - 1$ constants. These constants can be determined by conducting $3(2^{m-1}) - 1$ experiments, by assigning $3(2^{m-1}) - 1$ different sets of values to the resistances $(R_1, R_2, \ldots, R_m)$, and measuring the corresponding currents I_i. The obtained system of measurement equations has a unique solution for the constants if and only if $|\mathbf{M}| \neq 0$. If $|\mathbf{M}| = 0$, following the same strategy presented for the previous problems, one can determine the corresponding functional dependency. $\square$

2.2.4 Current Control Using Gyrator Resistance

In this problem we consider the design element to be the resistance of a gyrator. The gyrator resistance appears in the matrix $\mathbf{A}(\mathbf{p})$ with rank 2 dependency. We want to control the current I_i by a gyrator resistance, denoted by R_g, at an arbitrary location of the circuit.

Theorem 2.4 *In a linear DC circuit, the functional dependency of any current I_i on any gyrator resistance R_g can be determined by at most 5 measurements of the current I_i obtained for 5 different values of R_g.*

Proof. Let us consider the following two cases:

(1) the i-th branch is not connected to either port of the gyrator(Fig. 2.7a),
(2) the i-th branch is connected to one port of the gyrator (Fig. 2.7b).

Case 1: The i-th branch is not connected to either port of the gyrator

In this case, the matrices $\mathbf{B}_i(\mathbf{p}, \mathbf{q})$ and $\mathbf{A}(\mathbf{p})$, in (2.5), are both of rank 2 with respect to R_g. Therefore, according to Lemma 1.2, the functional dependency of I_i on R_g can be expressed as

$$I_i(R_g) = \frac{\tilde{\alpha}_0 + \tilde{\alpha}_1 R_g + \tilde{\alpha}_2 R_g^2}{\tilde{\beta}_0 + \tilde{\beta}_1 R_g + \tilde{\beta}_2 R_g^2}, \tag{2.34}$$

where $\tilde{\alpha}_0, \tilde{\alpha}_1, \tilde{\alpha}_2, \tilde{\beta}_0, \tilde{\beta}_1, \tilde{\beta}_2$ are constants. Assuming that $\tilde{\beta}_2 \neq 0$, one can divide the numerator and denominator of (2.34) by $\tilde{\beta}_2$ and obtain

$$I_i(R_g) = \frac{\alpha_0 + \alpha_1 R_g + \alpha_2 R_g^2}{\beta_0 + \beta_1 R_g + R_g^2}, \tag{2.35}$$

where $\alpha_0, \alpha_1, \alpha_2, \beta_0, \beta_1$ are constants. In order to determine these constants, one conducts 5 experiments by setting 5 different values to the gyrator resistance R_g, and measuring the corresponding currents I_i. In this case, the set of measurement equations will be

Fig. 2.7 An unknown linear DC circuit

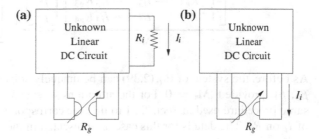

$$\begin{bmatrix} 1 & R_{g1} & R_{g1}^2 & -I_{i1} & -I_{i1}R_{g1} \\ 1 & R_{g2} & R_{g2}^2 & -I_{i2} & -I_{i2}R_{g2} \\ 1 & R_{g3} & R_{g3}^2 & -I_{i3} & -I_{i3}R_{g3} \\ 1 & R_{g4} & R_{g4}^2 & -I_{i4} & -I_{i4}R_{g4} \\ 1 & R_{g5} & R_{g5}^2 & -I_{i5} & -I_{i5}R_{g5} \end{bmatrix} \underbrace{\begin{bmatrix} \alpha_0 \\ \alpha_1 \\ \alpha_2 \\ \beta_0 \\ \beta_1 \end{bmatrix}}_{u} = \underbrace{\begin{bmatrix} I_{i1}R_{g1}^2 \\ I_{i2}R_{g2}^2 \\ I_{i3}R_{g3}^2 \\ I_{i4}R_{g4}^2 \\ I_{i5}R_{g5}^2 \end{bmatrix}}_{m}, \qquad (2.36)$$

$$\underbrace{\phantom{\begin{bmatrix} 1 & R_{g1} & R_{g1}^2 & -I_{i1} & -I_{i1}R_{g1} \end{bmatrix}}}_{M}$$

which has a unique solution for the constants α_0, α_1, β_0, β_1, β_2, if and only if $|M| \neq 0$ in (2.36). If $|M| = 0$ is the case, one can use the same procedure presented in Sect. 2.2.1 to derive the corresponding functional dependency of I_i on R_g. The details of this case are provided in the Appendix.

Case 2: The i-th branch is connected to one port of the gyrator

In this case, the matrix $B_i(p, q)$ is of rank 1 with respect to R_g; however, the matrix $A(p)$ is of rank 2 with respect to R_g. Therefore, using Lemma 1.2, the functional dependency of I_i on R_g can be written as

$$I_i(R_g) = \frac{\tilde{\alpha}_0 + \tilde{\alpha}_1 R_g}{\tilde{\beta}_0 + \tilde{\beta}_1 R_g + \tilde{\beta}_2 R_g^2}, \qquad (2.37)$$

where $\tilde{\alpha}_0, \tilde{\alpha}_1, \tilde{\beta}_0, \tilde{\beta}_1, \tilde{\beta}_2$ are constants. Supposing $\tilde{\beta}_2 \neq 0$ and dividing the numerator and denominator of (2.37) by $\tilde{\beta}_2$, one gets

$$I_i(R_g) = \frac{\alpha_0 + \alpha_1 R_g}{\beta_0 + \beta_1 R_g + R_g^2}, \qquad (2.38)$$

where $\alpha_0, \alpha_1, \beta_0, \beta_1$ are constants that can be determined by conducting 4 experiments, by assigning 4 different values to the gyrator resistance R_g, and measuring the corresponding currents I_i. Then, the following set of measurement equations can be formed

$$\underbrace{\begin{bmatrix} 1 & R_{g1} & -I_{i1} & -I_{i1}R_{g1} \\ 1 & R_{g2} & -I_{i2} & -I_{i2}R_{g2} \\ 1 & R_{g3} & -I_{i3} & -I_{i3}R_{g3} \\ 1 & R_{g4} & -I_{i4} & -I_{i4}R_{g4} \end{bmatrix}}_{M} \underbrace{\begin{bmatrix} \alpha_0 \\ \alpha_1 \\ \beta_0 \\ \beta_1 \end{bmatrix}}_{u} = \underbrace{\begin{bmatrix} I_{i1}R_{g1}^2 \\ I_{i2}R_{g2}^2 \\ I_{i3}R_{g3}^2 \\ I_{i4}R_{g4}^2 \end{bmatrix}}_{m}. \qquad (2.39)$$

As before, the system of Eq. (2.39) can be uniquely solved for the constants α_0, α_1, β_0, β_1, provided $|M| \neq 0$. For the situations where $|M| = 0$, one can follow the same procedure used in Sect. 2.2.1 to find the corresponding functional dependency of I_i on R_g. The details for this case are presented in the Appendix.   □

Fig. 2.8 An unknown linear
DC circuit

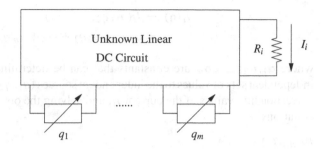

2.2.5 Current Control Using m Independent Sources

Here, we consider the problem of controlling the current in the i-th branch of an
unknown linear DC circuit, denoted by I_i, by only using the independent cur-
rent/voltage sources, denoted by $\mathbf{q} = [q_1, q_2, \ldots, q_m]^T$, at arbitrary locations of
the circuit (Fig. 2.8).

Theorem 2.5 *In a linear DC circuit, the functional dependency of any current I_i
on the independent sources can be determined by m measurements of the current I_i
obtained for m linearly independent sets of values of the source vector* $\mathbf{q}$.

Proof. Recall (2.4),

$$\mathbf{b}(\mathbf{q}) = q_1\mathbf{b}_1 + q_2\mathbf{b}_2 + \cdots + q_m\mathbf{b}_m, \tag{2.40}$$

where $q_1, q_2, \ldots, q_m$ represent the independent sources. The matrix $\mathbf{B}_i(\mathbf{p}, \mathbf{q})$ in (2.5)
can be written as

$$\mathbf{B}_i(\mathbf{p}, \mathbf{q}) = [\mathbf{A}_1(\mathbf{p}), \ldots, \mathbf{A}_{i-1}(\mathbf{p}), \mathbf{b}(\mathbf{q}), \mathbf{A}_{i+1}(\mathbf{p}), \ldots, \mathbf{A}_n(\mathbf{p})]. \tag{2.41}$$

Therefore, the matrix $\mathbf{B}_i(\mathbf{p}, \mathbf{q})$ is of rank 1 with respect to each of the independent
sources $q_1, q_2, \ldots, q_m$ and $|\mathbf{B}_i(\mathbf{p}, \mathbf{q})|$ can be written as a linear combination of the
parameters $q_1, q_2, \ldots, q_m$,

$$|\mathbf{B}_i(\mathbf{p}, \mathbf{q})| = q_1|\mathbf{B}_{i1}(\mathbf{p})| + q_2|\mathbf{B}_{i2}(\mathbf{p})| + \cdots + q_m|\mathbf{B}_{im}(\mathbf{p})|, \tag{2.42}$$

where

$$\mathbf{B}_{ij}(\mathbf{p}) = [\mathbf{A}_1(\mathbf{p}), \ldots, \mathbf{A}_{i-1}(\mathbf{p}), \mathbf{b}_j, \mathbf{A}_{i+1}(\mathbf{p}), \ldots, \mathbf{A}_n(\mathbf{p})], \tag{2.43}$$

for $j = 1, 2, \ldots, m$. The matrix $\mathbf{A}(\mathbf{p})$ is of rank 0 with respect to $q_1, q_2, \ldots, q_m$ and
thus, according to Lemma 1.1, $|\mathbf{A}(\mathbf{p})|$ is a constant. Hence, the functional dependency
of I_i on $q_1, q_2, \ldots, q_m$ can be expressed as

$$I_i(\mathbf{q}) := I_i(q_1, q_2, \ldots, q_m)$$
$$= \alpha_1 q_1 + \alpha_2 q_2 + \cdots + \alpha_m q_m, \tag{2.44}$$

where $\alpha_1, \alpha_2, \ldots, \alpha_m$ are constants that can be determined by setting m linearly independent sets of values to the independent sources $(q_1, q_2, \ldots, q_m)$, measuring the corresponding values of the current I_i, and solving the obtained set of measurement equations. $\square$

Remark 2.3

1. Theorem 2.5 is the well-known Superposition Principle of circuit theory.
2. If the independent sources vary in the intervals $q_j^- \leq q_j \leq q_j^+$, $j = 1, 2, \ldots, m$, then the current I_i will vary in an interval whose end values can be computed using the vertices (q_j^-, q_j^+), $j = 1, 2, \ldots, m$. For example suppose that I_i is given as below,

$$I_i(\mathbf{q}) = 2q_1 - q_2 + 5q_3 - 3q_4, \tag{2.45}$$

where $q_j^- \leq q_j \leq q_j^+$, $q_j^- \geq 0$, $j = 1, 2, 3, 4$. One may decompose I_i as

$$I_i(\mathbf{q}) = 2q_1 - q_2 + 5q_3 - 3q_4 = (2q_1 + 5q_3) - (q_2 + 3q_4). \tag{2.46}$$

Then, the maximum and minimum values of I_i, denoted by $I_i^{\max}$ and $I_i^{\min}$, respectively, can be obtained from

$$I_i^{\max} = (2q_1^+ + 5q_3^+) - (q_2^- + 3q_4^-), \tag{2.47}$$
$$I_i^{\min} = (2q_1^- + 5q_3^-) - (q_2^+ + 3q_4^+). \tag{2.48}$$

2.3 Power Level Control

In this section we consider another class of circuit synthesis problems where, in an unknown linear DC circuit, the power level in a resistor is to be controlled by adjusting the design elements at arbitrary locations of the circuit. For the sake of simplicity, suppose that the resistor R_i is located in the i-th branch of the circuit and we wish to control the power level P_i, in the resistor R_i, by some design elements. As in the previous section, we consider several cases of design elements and provide the results for each case.

2.3.1 Power Level Control Using a Single Resistor

In this subsection we show how to control the power level P_i in the resistor R_i, located in the i-th branch of an unknown linear DC circuit, by adjusting any resistor R_j at an arbitrary location of the circuit. Assume that R_j is not a gyrator resistance and recall the results developed in Sect. 2.2.1.

Theorem 2.6 *In a linear DC circuit, the functional dependency of the power level P_i, in the resistor R_i, on any resistance R_j can be determined by at most 3 measurements of the current I_i (passing through R_i) obtained for 3 different values of R_j, and 1 measurement of the voltage across the resistor R_i, corresponding to one of the resistance settings.*

Proof. Let us consider two cases: (1) $i \neq j$, and (2) $i = j$.

Case 1: $i \neq j$

We can write the power level P_i as $P_i = \frac{V_i}{I_i} I_i^2$. The functional dependency of the current I_i, passing through R_i, on any other resistance R_j will be of either forms (2.8) or (2.10). Since the ratio $\frac{V_i}{I_i}$ is the same for each experiment, then only one measurement of the voltage V_i, across the resistor R_i, in addition to the 3 measurements of the current I_i, is required to determine the functional dependency of P_i on R_j. Assuming one measures V_{i1} from the first experiment, then the functional dependency of P_i on R_j can be expressed as one of the following forms:

- If $|\mathbf{M}| \neq 0$ in (2.9):

$$P_i(R_j) = \frac{V_{i1}}{I_{i1}} \left(\frac{\alpha_0 + \alpha_1 R_j}{\beta_0 + R_j} \right)^2, \qquad (2.49)$$

where V_{i1} and I_{i1} are the voltage and current signals, at the resistor R_i, measured from the first experiment, and the constants $\alpha_0, \alpha_1, \beta_0$ are obtained by solving (2.9), as explained in Sect. 2.2.1.
- If $|\mathbf{M}| = 0$ in (2.9):

$$P_i(R_j) = \frac{V_{i1}}{I_{i1}} (\alpha_0 + \alpha_1 R_j)^2, \qquad (2.50)$$

where V_{i1} and I_{i1} are the voltage and current signals, at the resistor R_i, measured from the first experiment, and the constants α_0, α_1 can be determined using any two of the conducted experiments, as discussed in Sect. 2.2.1.

Case 2: $i = j$

Let us write the power level P_i as $P_i = R_i I_i^2$. Based on the results of Sect. 2.2.1, the functional dependency of I_i on R_i will be of either forms given in (2.12) or (2.14). Hence, the functional dependency of P_i on R_i will be of one the following forms:

- If $|\mathbf{M}| \neq 0$ in (2.13):

$$P_i(R_i) = R_i \left(\frac{\alpha_0}{\beta_0 + R_i} \right)^2, \qquad (2.51)$$

where the constants α_0, β_0 can be obtained as explained in Sect. 2.2.1.
- If $|\mathbf{M}| = 0$ in (2.13):

$$P_i(R_i) = \alpha_0^2 R_i, \tag{2.52}$$

where α_0 is a constant that can be determined as discussed in Sect. 2.2.1.   □

Remark 2.4 Suppose that $i = j$ and $|\mathbf{M}| \neq 0$ in (2.13), then the derivative of P_i, in (2.51), with respect to R_i, can be calculated as

$$\frac{dP_i}{dR_i} = \frac{\alpha_0^2(\beta_0 - R_i)}{(\beta_0 + R_i)^3}. \tag{2.53}$$

We have the following statements:

1. The functional dependency of P_i on R_i, in (2.51), in this case, is *not* monotonic. As $R_i \to 0$, $P_i \to 0$ and when $R_i \to \infty$, $P_i \to 0$. Therefore, as the value of the resistance R_i increases from 0 to ∞, the power P_i increases from 0 to the maximum achievable value of $\frac{\alpha_0^2}{4\beta_0}$, and then decreases to 0 at very large values of R_i. The maximum occurs at $R_i = \beta_0$.
2. The achievable range for the power level P_i, by varying the resistance R_i in the interval $[0, \infty)$, is

$$0 \leq P_i < \frac{\alpha_0^2}{4\beta_0}. \tag{2.54}$$

3. In a power level control problem of this type, for any desired prescribed interval of power P_i, which is within the achievable range (2.54), one may find *two* ranges of values for the design resistance R_i.

2.3.2 Power Level Control Using Two Resistors

In this case it is desired to control the power level P_i, by adjusting any two resistors R_j and R_k at arbitrary locations of the circuit. Assuming that R_j and R_k are not gyrator resistances, and based on the results of Sect. 2.2.2, we have the following theorem.

Theorem 2.7 *In a linear DC circuit, the functional dependency of the power level P_i, in any resistor R_i, on any two resistances R_j and R_k can be determined by at most 7 measurements of the currents I_i (passing through R_i) obtained for 7 different sets of values (R_j, R_k), and 1 measurement of the voltage across the resistor R_i, corresponding to one of the resistance settings.*

Proof. The proof is similar to the previous case and thus omitted here.   □

In this problem, the power level P_i can be depicted as a surface in a 3D plot. In a synthesis problem of this type, any constraint on the power level P_i results in a corresponding region in the R_j–R_k plane, if the solution set to that constraint is not empty.

Remark 2.5 For the case of m resistors, the functional dependencies can be derived similarly using the results given in Sect. 2.2.3.

2.3.3 Power Level Control Using Gyrator Resistance

Here, we want to control the power level P_i using any gyrator resistance R_g, at an arbitrary location of the circuit. The functional dependency of the current I_i on any gyrator resistance R_g is obtained in Sect. 2.2.4. Applying the same technique, one can find the functional dependency of P_i on any gyrator resistance R_g. We leave the details to the reader.

2.4 Examples of DC Circuit Design

Example 2.1. In this example we show how the method proposed in Sects. 2.2.1 and 2.3.1 can be used toward control design problems in unknown linear DC circuits. Consider the unknown circuit shown in Fig. 2.9.

In this example, it is desired to find the functional dependency of the current I_1 on the resistance R_9. Based on the results given in Sect. 2.2.1, one conducts 3 experiments, by setting 3 different values to R_9, and measuring the corresponding currents I_1. Suppose that experiments are done and let Table 2.1 summarize the numerical values, for this example, obtained from the 3 experiments.

Substituting the numerical values obtained from the experiments into the matrix $\mathbf{M}$ in (2.9) resulted in $|\mathbf{M}| \neq 0$. Therefore, (2.9) can be uniquely solved for the constants

Fig. 2.9 An unknown resistive circuit

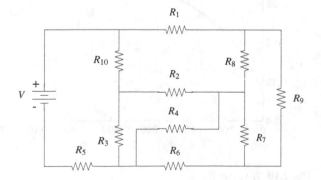

Exp. no.	R_9 (Ω)	I_1 (A)
Table 2.1 Numerical values of the measurements for the DC circuit Example 2.1		
1	1	0.054
2	5	0.056
3	10	0.058

and yield the following functional dependency which is plotted in Fig. 2.10.

$$I_1(R_9) = \frac{78.4 + 0.66R_9}{181.3 + R_9}. \tag{2.55}$$

Remark 2.6

1. The current I_1 monotonically increases as R_9 increases.
2. By varying R_9 in the range $[0, \infty)$, the achievable range for I_1 becomes $[\frac{\alpha_0}{\beta_0}, \alpha_1] =$ [0.43, 0.66].
3. In a synthesis problem where the current I_1 is to be controlled to stay within an acceptable interval, since I_1 is monotonic in R_9, one can find a corresponding interval for R_9 values for which the current I_1 stays within the acceptable range.

Suppose that we wish to design R_9 such that I_1 lies within the following achievable range

$$0.5 \leq I_1 \leq 0.6 \ (A). \tag{2.56}$$

Using (2.55), or Fig. 2.10, the corresponding range for the design resistor R_9 can be obtained as

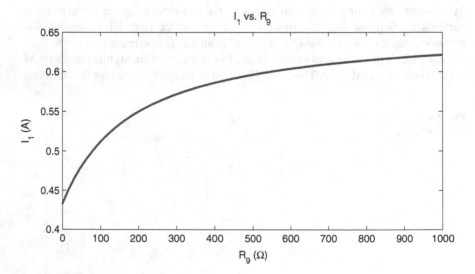

Fig. 2.10 I_1 versus R_9

Table 2.2 Numerical values of the measurements for the DC circuit Example 2.2

Exp. no.	R_9 (Ω)	I_1 (A)	I_3 (A)	I_9 (A)
1	1	0.437	0.964	0.301
2	5	0.438	0.972	0.295
3	10	0.444	0.982	0.287
Exp. no.	R_9 (Ω)	V_1 (V)	V_3 (V)	
1	1	8.67	4.82	

$$79 \leq R_9 \leq 550 \ (\Omega). \tag{2.57}$$

Example 2.2. Consider the same circuit as above (Fig. 2.9). Suppose now that the power levels within R_1, R_3 and R_9, denoted by P_1, P_3 and P_9, respectively, must remain in the following ranges:

$$6 \ (W) \leq P_1 \leq 7 \ (W), \tag{2.58}$$

$$7 \ (W) \leq P_3 \leq 8 \ (W), \tag{2.59}$$

$$3 \ (W) \leq P_9 \leq 3.5 \ (W). \tag{2.60}$$

Assume that the design resistor is R_9. Based on the results of Sect. 2.3.1, one conducts 3 experiments by assigning 3 different values to R_9, and measuring the corresponding currents I_1, I_3 and I_9, passing through the resistors R_1, R_3 and R_9, respectively. In this problem, one also needs to measure the voltage across R_1 and R_3 from one of the experiments. Suppose that the experiments are done and let Table 2.2 summarize the numerical values for this example, obtained from the experiments.

Substituting the numerical values from Table 2.2 into the matrix **M** in (2.9), for the currents I_1 and I_3, and into the matrix **M** in (2.13), for the current I_9, yields $|\mathbf{M}| \neq 0$, for all cases. Therefore, the functional dependencies of P_1, P_3 and P_9 on R_9 will be

$$P_1(R_9) = \frac{8.67}{0.437} \left(\frac{78.4 + 0.66 R_9}{181.3 + R_9} \right)^2, \tag{2.61}$$

$$P_3(R_9) = \frac{4.82}{0.964} \left(\frac{174.4 + 1.34 R_9}{181.3 + R_9} \right)^2, \tag{2.62}$$

$$P_9(R_9) = R_9 \left(\frac{54.9}{181.3 + R_9} \right)^2. \tag{2.63}$$

Figure 2.11 shows the plots of the power levels P_1, P_3 and P_9 obtained above.

Using (2.61)–(2.63), shown graphically in Fig. 2.11, one imposes the power level constraints (2.58)–(2.60) to find the corresponding ranges of R_9 values. A necessary condition for the existence of a solution is that the constraints (2.58)–(2.60) must be within their corresponding achievable ranges. For this example, the power level constraints are within the achievable ranges; hence, we can find the following ranges for R_9 values:

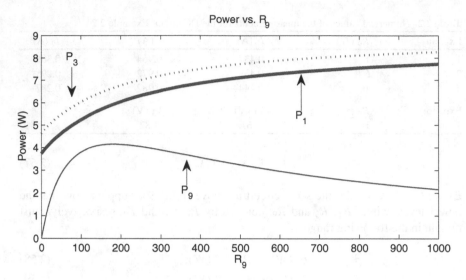

Fig. 2.11 P_1, P_3, P_9 versus R_9

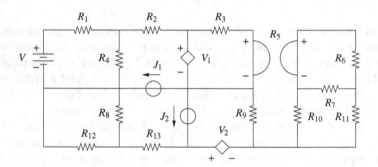

Fig. 2.12 An unknown linear DC circuit

$$190 \; (\Omega) \leq R_9 \leq 450 \; (\Omega), \tag{2.64}$$

$$250 \; (\Omega) \leq R_9 \leq 690 \; (\Omega), \tag{2.65}$$

$$60 \; (\Omega) \leq R_9 \leq 80 \; \cup \; 420 \; (\Omega) \leq R_9 \leq 580 \; (\Omega), \tag{2.66}$$

corresponding to the power level constraints (2.58), (2.59) and (2.60), respectively. Therefore, the range for R_9 values where (2.58), (2.59) and (2.60) are achieved simultaneously is the intersection of the ranges calculated above, that is

$$420 \; (\Omega) \leq R_9 \leq 450 \; (\Omega). \tag{2.67}$$

Example 2.3. Consider the unknown linear DC circuit shown in Fig. 2.12.

In this example, R_i, $i = 1, 2, \ldots, 13$, $i \neq 5$ are resistors, R_5 is a gyrator resistance, V, J_1, J_2 are independent sources and V_1, V_2 are dependent sources. Our goal is to control the power levels in R_3, R_6 and R_{11}, denoted by P_3, P_6 and P_{11}, respectively, to be within the following ranges:

$$40 \ (W) \leq P_3 \leq 60 \ (W), \tag{2.68}$$

$$1 \ (W) \leq P_6 \leq 8 \ (W), \tag{2.69}$$

$$0.5 \ (W) \leq P_{11} \leq 5 \ (W). \tag{2.70}$$

Assume that the design elements are the resistances R_1 and R_6. Therefore, we need to find the region in the R_1–R_6 plane where the constraints (2.68), (2.69) and (2.70) are satisfied. Based on the approach presented in Sect. 2.3.2, in order to find the functional dependency of any power level in terms of any two resistances, one needs to do at most 7 measurements of current and one measurement of voltage. Let us treat each power level problem separately as follows:

(a) P_3 versus R_1 and R_6
Based on the results obtained in Sect. 2.3.2, in order to find the functional dependency of P_3 on R_1 and R_6, one needs to conduct 7 experiments by setting 7 different sets of values for the resistances (R_1, R_6), and measuring the corresponding values for current I_3. In addition to the current measurements, one measurement of the voltage, across the resistor R_3, is needed to determine the functional dependency of interest. Suppose that this measurement is taken from the first experiment and denoted as V_{31}. Suppose that experiments are done and let Table 2.3 summarize the numerical values assigned to the resistances R_1 and R_6 along with the corresponding measurements of I_3 and V_{31}. Substituting the numerical values of Table 2.3 into the matrix $\mathbf{M}$, in (2.26), it can be verified that $|\mathbf{M}| \neq 0$. Thus, the functional dependency of interest will be of the form

$$P_3(R_1, R_6) = \frac{V_{31}}{I_{31}} \underbrace{\left(\frac{\alpha_0 + \alpha_1 R_1 + \alpha_2 R_6 + \alpha_3 R_1 R_6}{\beta_0 + \beta_1 R_1 + \beta_2 R_6 + R_1 R_6} \right)^2}_{I_3^2(R_1, R_6)}, \tag{2.71}$$

Table 2.3 Numerical values of the measurements for the DC circuit Example 2.3

Exp. no.	$R_1(\Omega)$	$R_6(\Omega)$	$I_3(A)$
1	7	1	3.33
2	13	8	2.71
3	21	19	2.47
4	35	26	2.57
5	40	32	2.52
6	52	45	2.47
7	59	56	2.44
Exp. no.	$R_1(\Omega)$	$R_6(\Omega)$	$V_{31}(V)$
1	7	1	33.3

(a) 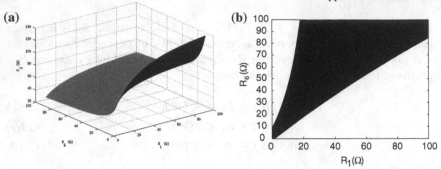 **(b)**

Fig. 2.13 **a** P_3 versus R_1 and R_6. **b** Region (in *black color*) where (2.68) is satisfied

(a) 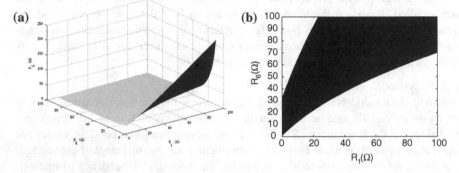 **(b)**

Fig. 2.14 **a** P_6 versus R_1 and R_6. **b** Region (in *black color*) where (2.69) is satisfied

where the constants $\alpha_0, \alpha_1, \alpha_2, \alpha_3, \beta_0, \beta_1, \beta_2$ can be determined by solving (2.26), using the numerical values of Table 2.3. For this example, the constants are obtained as: $\alpha_0 = 98.4$, $\alpha_1 = 36$, $\alpha_2 = 6.6$, $\alpha_3 = 2.4$, $\beta_0 = 58.5$, $\beta_1 = 5$, $\beta_2 = 11.7$. Hence, the functional dependency of P_3 on R_1 and R_6 will be

$$P_3(R_1, R_6) = \frac{33.3}{3.33} \left(\frac{98.4 + 36R_1 + 6.6R_6 + 2.4R_1R_6}{58.5 + 5R_1 + 11.7R_6 + R_1R_6} \right)^2. \tag{2.72}$$

Figure 2.13(a) shows the plot of the surface P_3 as a function of the design elements R_1 and R_6, obtained in (2.72). Applying constraint (2.68) on P_3, one may obtain the region in the R_1–R_6 plane, shown in black color in Fig. 2.13(b), where this constraint is satisfied.

(b) P_6 versus R_1 and R_6

The functional dependency of P_6 on R_1 and R_6 can be determined by at most 5 measurements of current and one measurement of voltage as discussed in Sect. 2.3.2 (Case 2). The plot of the surface P_6 as a function of R_1 and R_6 is shown in Fig. 2.14(a). Applying constraint (2.69) on P_6, one may obtain the region in the R_1–R_6 plane, shown in black color in Fig. 2.14(b), where this constraint is valid.

(c) P_{11} versus R_1 and R_6

(a)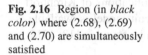

(b)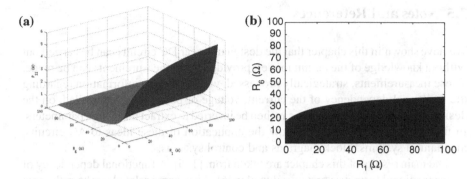

Fig. 2.15 a P_{11} versus R_1 and R_6. **b** Region (in *black color*) where (2.70) is satisfied

Fig. 2.16 Region (in *black color*) where (2.68), (2.69) and (2.70) are simultaneously satisfied

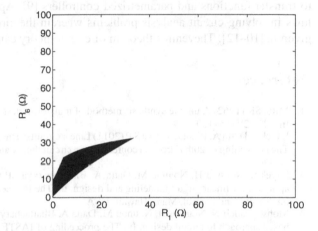

Following the same procedure used to determine the functional dependency of P_3 on R_1 and R_6, one can determine the dependency of P_{11} on R_1 and R_6. The plot of the surface P_{11} as a function of R_1 and R_6 is shown in Fig. 2.15(a). Applying constraint (2.70) on P_{11}, one finds the region in the R_1–R_6 plane, shown in black color in Fig. 2.15(b), where this constraint is satisfied.

In order to satisfy the constraints given in (2.68), (2.69) and (2.70), simultaneously, one has to intersect the regions shown in Figs. 2.13(b), 2.14(b), 2.15(b). Figure 2.16 shows the region (in black color) in the R_1–R_6 plane where constraints (2.68), (2.69) and (2.70) are satisfied, simultaneously.

2.5 Notes and References

We have shown in this chapter that the design of linear DC circuits can be carried out without knowledge of the circuit model, provided a few measurements can be made. These measurements, strategically processed, yield complete information regarding the functional dependency of the current, voltage or power to be controlled, on the design variables. These relations can then be inverted to extract the design parameters. In the subsequent chapters we show the application of these ideas to AC circuits, mechanical systems, block diagrams and control systems.

The main results of this chapter are taken from [1–5]. A functional dependency of linear fractional form was first introduced in [6]. Also, some related works in the area of symbolic network functions can be found in [7, 8]. This approach can be extended to transfer functions and parametrized controllers [9]. Applications of Kirchhoff's laws in solving circuit analysis problems wherein the circuit model is available is given in [10–12]. Thevenin's theorem of circuit theory can be found in [13–15].

References

1. Mitra SK (1962) A unique synthesis method of transformerless active rc networks. J Franklin Inst 274(2):115–129
2. Layek R, Datta A, Bhattacharyya S P (2011) Linear circuits: a measurement based approach. In: The proceeding of 20th European conference on circuit theory and design, Linkoping, Sweden, pp 476–479
3. Layek R, Nounou H, Nounou M, Datta A, Bhattacharyya SP (2012) A measurement based approach for linear circuit modeling and design. In: The Proceeding of 51th IEEE conference on decision and control, Maui, Hawaii, USA
4. Mohsenizadeh N, Nounou H, Nounou M, Datta A, Bhattacharyya SP (2012) A measurement based approach to circuit design. In: The proceeding of IASTED international conference on engineering and applied science, Colombo, Sri Lanka, pp 27–34
5. Bhattacharyya SP (2013) Linear sytems: a measurement based approach to analysis, synthesis and design. In: The 3rd IASTED Asian conference on modelling. Identification and control, Phuket, Thailand
6. DeCarlo R, Lin P (1995) Linear circuit analysis: time domain, phasor, and laplace transform approaches. Prentice Hall, Englewood Cliffs, NJ
7. Lin P (1973) A survey of applications of symbolic network functions. IEEE Trans Circuit Theory 20(6):732–737
8. Alderson GE, Lin P (1973) Computer generation of symbolic network functions-a new theory and implementation. IEEE Trans Circuit Theory 20(1):48–56
9. Hara S (1987) Parametrization of stabilizing controllers for multivariable servo systems with two degrees of freedom. Int J Control 45(3):779–790
10. Kirchhoff G (1847) Ueber die auflsung der gleichungen, auf welche man bei der untersuchung der linearen vertheilung galvanischer strme gefhrt wird. Annalen der Physik 148(12):497–508
11. Kailath T (1980) Linear systems. Prentice-Hall, Englewood Cliffs, NJ
12. Varaiya P (2002) Structure and interpretation of signals and systems. Addison-Wesley, Boston, MA
13. Thévenin L (1883) Sur un nouveau théoréme d'électricité dynamique [on a new theorem of dynamic electricity]. C. R. des Séances de l'Académie des Sciences 97:159–161
14. Brittain J (1990) Thévenin's theorem. IEEE Spectrum 27(3):42
15. Johnson DH (April 2003) Origins of the equivalent circuit concept: the voltage-source equivalent. Proc IEEE 91(4):636–640

Chapter 3
Application to AC Circuits

In this chapter we extend our measurement based approach to the domain of AC circuits operating in steady state at a fixed frequency. The main difference between the approach presented here for AC circuits and its DC circuit counterpart (Chap. 2) is that in an AC circuit the signals and variables are complex quantities usually called phasors and impedances, rather than real quantities.

3.1 Current Control

In this section we consider synthesis problems in unknown AC circuits, operating in steady state at a fixed frequency. The current, voltage or power phasor in any branch of the circuit is to be controlled by adjusting the design elements, typically impedances, at arbitrary locations. We consider several cases of design elements and provide the results for each case.

The governing steady state equations of a linear AC circuit, operating at a fixed frequency ω, can be represented in the following matrix form

$$\mathbf{A}(\mathbf{p}(j\omega))\mathbf{x}(j\omega) = \mathbf{b}(\mathbf{q}(j\omega)), \qquad (3.1)$$

where $\mathbf{A}(\mathbf{p}(j\omega))$ is the circuit characteristic matrix containing the circuit impedances, $\mathbf{x}(j\omega)$ is the vector of unknown current phasors and $\mathbf{q}(j\omega)$ represents the vector of independent voltage and current sources. Suppose that we want to control the current phasor in the i-th branch of the circuit, denoted by $I_i(j\omega)$. Applying Cramer's rule to (3.1), $I_i(j\omega)$ can be calculated from

$$x_i(j\omega) = I_i(j\omega) = \frac{|\mathbf{B}_i(\mathbf{p}(j\omega), \mathbf{q}(j\omega))|}{|\mathbf{A}(\mathbf{p}(j\omega))|}, \qquad (3.2)$$

where $\mathbf{B}_i(\mathbf{p}(j\omega), \mathbf{q}(j\omega))$ is the matrix obtained by replacing the i-th column of the characteristic matrix $\mathbf{A}(\mathbf{p}(j\omega))$ by the vector $\mathbf{b}(\mathbf{q}(j\omega))$. We assume that the circuit

S. P. Bhattacharyya et al., *Linear Systems*, SpringerBriefs in Applied Sciences
and Technology, DOI: 10.1007/978-81-322-1641-4_3, © The Author(s) 2014

is unknown, so that the matrices $\mathbf{B}_i(\mathbf{p}(j\omega), \mathbf{q}(j\omega))$ and $\mathbf{A}(\mathbf{p}(j\omega))$ are unknown. In the following subsections, for each case of the design elements, a general rational function for the current phasor $I_i(\omega)$, in terms of the design elements, will be derived. For the sake of simplicity, we drop the argument $(j\omega)$ in writing the equations from now on.

3.1.1 Current Control Using a Single Impedance

Consider the unknown linear AC circuit shown in Fig. 3.1. We want to control the current phasor in the i-th branch, I_i, by adjusting any impedance Z_j at an arbitrary location of the circuit. Assume that Z_j is not a gyrator resistance.

Theorem 3.1 *In a linear AC circuit, the functional dependency of any current phasor I_i on any impedance Z_j can be determined by at most three measurements of the current phasor I_i obtained for three different complex values of Z_j.*

Proof The proof is similar to its D.C. circuit counterpart provided in Sect. 2.2.1. The main difference is that the circuit signals/variables and the constants appearing in the functional dependencies will be complex quantities rather than real numbers. Therefore, we provide the results and leave the details to the reader.

Case 1: $i \neq j$

The functional dependency of I_i on Z_j will be of the form

$$I_i(Z_j) = \frac{\alpha_0 + \alpha_1 Z_j}{\beta_0 + Z_j}, \tag{3.3}$$

where $\alpha_0, \alpha_1, \beta_0$ are complex quantities that can be uniquely determined by solving the measurement equations,

Fig. 3.1 An unknown linear AC circuit

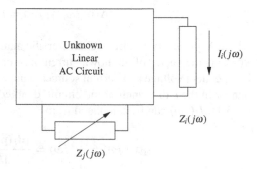

$$\underbrace{\begin{bmatrix} 1 & Z_{j1} & -I_{i1} \\ 1 & Z_{j2} & -I_{i2} \\ 1 & Z_{j3} & -I_{i3} \end{bmatrix}}_{\mathbf{M}} \underbrace{\begin{bmatrix} \alpha_0 \\ \alpha_1 \\ \beta_0 \end{bmatrix}}_{\mathbf{u}} = \underbrace{\begin{bmatrix} Z_{j1}I_{i1} \\ Z_{j2}I_{i2} \\ Z_{j3}I_{i3} \end{bmatrix}}_{\mathbf{m}}, \tag{3.4}$$

provided $|\mathbf{M}| \neq 0$. These complex quantities can be written as

$$\alpha_0(j\omega) = \alpha_{0r}(\omega) + j\alpha_{0i}(\omega),$$
$$\alpha_1(j\omega) = \alpha_{1r}(\omega) + j\alpha_{1i}(\omega),$$
$$\beta_0(j\omega) = \beta_{0r}(\omega) + j\beta_{0i}(\omega).$$

If $|\mathbf{M}| = 0$ in (3.4), the functional dependency of I_i on Z_j can be expressed as

$$I_i(Z_j) = \alpha_0 + \alpha_1 Z_j, \tag{3.5}$$

where the complex quantities α_0, α_1 can be determined from any two of the experiments conducted earlier.

Case 2: $i = j$

In this case, the functional dependency of I_i on Z_i can be represented as

$$I_i(Z_i) = \frac{\alpha_0}{\beta_0 + Z_i}, \tag{3.6}$$

where α_0, β_0 are complex quantities that can be calculated by solving the measurement equations,

$$\underbrace{\begin{bmatrix} 1 & -I_{i1} \\ 1 & -I_{i2} \end{bmatrix}}_{\mathbf{M}} \underbrace{\begin{bmatrix} \alpha_0 \\ \beta_0 \end{bmatrix}}_{\mathbf{u}} = \underbrace{\begin{bmatrix} I_{i1}Z_{i1} \\ I_{i2}Z_{i2} \end{bmatrix}}_{\mathbf{m}}. \tag{3.7}$$

if and only if $|\mathbf{M}| \neq 0$. In a situation where $|\mathbf{M}| = 0$ in (3.7), the current phasor I_i will be a constant,

$$I_i(Z_i) = \alpha_0, \tag{3.8}$$

where the complex quantity α_0 can be obtain from one of the experiments conducted earlier. □

As noted earlier, since the main difference between the results of this chapter and their D.C. circuit counterparts is that in AC circuits the variables are complex quantities, rather than real numbers, the proofs of the following theorems are omitted.

Fig. 3.2 An unknown linear
AC circuit

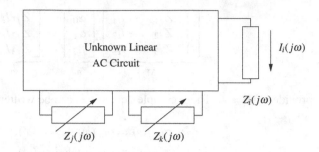

3.1.2 Current Control Using Two Impedances

Consider the unknown linear AC circuit shown in Fig. 3.2. In this problem, we want to control the current phasor in the i-th branch, I_i, by adjusting any two impedances Z_j and Z_k at arbitrary locations of the circuit. Assume that Z_j and Z_k are not gyrator resistances.

Theorem 3.2 *In a linear AC circuit, the functional dependency of any current phasor I_i on any two impedances Z_j and Z_k can be determined by at most seven measurements of the current phasor I_i obtained for seven different sets of complex values (Z_j, Z_k).*

Remark 3.1. If m impedances are considered as the design elements, one can use the results provided in Sect. 2.2.3 to derive the corresponding functional dependencies.

3.1.3 Current Control Using Gyrator Resistance

In this problem the design element is the resistance of a gyrator; thus, our objective is to control the current phasor I_i by a gyrator resistance, denoted by R_g, at an arbitrary location of the circuit.

Theorem 3.3 *In a linear AC circuit, the functional dependency of any current phasor I_i on any gyrator resistance R_g can be determined by at most seven measurements of the current phasor I_i obtained for five different values of R_g.*

3.1.4 Current Control Using m Independent Sources

Consider the problem of controlling the current phasor I_i using the independent current and voltage sources, denoted by $q_1, q_2, \ldots, q_m$, at arbitrary locations of the circuit (Fig. 3.3).

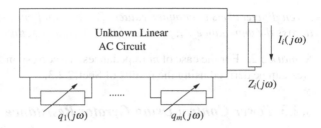

Fig. 3.3 An unknown linear AC circuit

Theorem 3.4 *In a linear AC circuit, the functional dependency of any current phasor I_i on the independent sources can be determined by m measurements of the current phasor I_i obtained for m linearly independent sets of values of the source vector $\mathbf{q} = [q_1, q_2, \ldots, q_m]^T$.*

3.2 Power Control

3.2.1 Power Control Using a Single Impedance

In this problem, our objective is to control the complex power P_i, in the impedance Z_i, located in the i-th branch of an unknown linear AC circuit, by adjusting any impedance Z_j at an arbitrary location of the circuit. Assuming that Z_j is not a gyrator resistance and recalling the results presented in Sect. 3.1.1, we have the following theorem.

Theorem 3.5 *In a linear AC circuit, the functional dependency of the complex power P_i on any impedance Z_j can be determined by at most three measurements of the current phasor I_i (passing through Z_i) obtained for three different complex values of Z_j, and one measurement of the voltage across the impedance Z_i, for one such setting of the impedance.*

3.2.2 Power Control Using Two Impedances

Suppose that the power P_i is to be controlled by adjusting two impedances Z_j and Z_k at arbitrary locations of the circuit. Assume that Z_j and Z_k are not gyrator resistances. Based on the results developed in Sect. 3.1.2, we can state the following theorem.

Theorem 3.6 *In a linear AC circuit, the functional dependency of the power level P_i, in any impedance Z_i, on any two impedances Z_j and Z_k can be determined by at most seven measurements of the current phasor I_i (passing through Z_i) obtained for* *seven different sets of complex values* (Z_j, Z_k), *and one measurement of the voltage across the impedance* Z_i, *for one of the impedance settings.*

Remark 3.2. For the case of m impedances, the corresponding functional dependencies can be derived using the results of Sect. 2.2.3.

3.2.3 Power Control Using Gyrator Resistance

In this case, the power P_i is to be controlled by a gyrator resistance R_g, at an arbitrary location of the circuit. Using the results obtained in Sect. 3.1.3, we have the following theorem.

Theorem 3.7 *In a linear AC circuit, the functional dependency of the complex power P_i on any gyrator resistance R_g can be determined by at most five measurements of the current phasor I_i obtained for five different values of R_g, and one measurement of the voltage across the impedance Z_i, corresponding to one of the impedance settings.*

3.3 An Example of AC Circuit Design

This example shows how the approach developed for the AC circuits can be used to solve a synthesis problem. Consider the unknown linear AC circuit shown in Fig. 3.4.

Let us assume that the operating frequency of this AC circuit is $f = 60$ (Hz) and thus the angular frequency will be $\omega_0 = 2\pi f = 120\pi$. Suppose that we want to control the current phasors I_3 and I_9 to be within the following ranges:

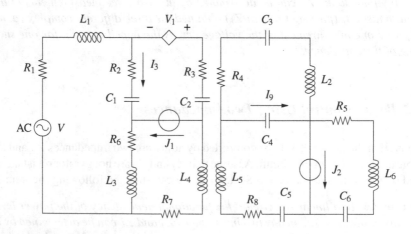

Fig. 3.4 An unknown linear AC circuit

$$0\ (A) \le |I_3| \le 4\ (A), \tag{3.9}$$

$$10\ (deg) \le \angle I_3 \le 30\ (deg), \tag{3.10}$$

$$0\ (A) \le |I_9| \le 2.5\ (A), \tag{3.11}$$

$$-30\ (deg) \le \angle I_9 \le -10\ (deg). \tag{3.12}$$

Assume that the design elements are the inductor L_1 and the capacitor C_2. Therefore, we need to find the region in the L_1–C_2 plane where the constraints (3.9)–(3.12) are satisfied. Based on the approach provided in Sect. 3.1.2, one can determine the functional dependency of any current phasor in terms of any two impedances by taking at most seven measurements of the current phasor. Let us treat each current phasor problem separately as follows:

(a) I_3 versus L_1 and C_2

To determine the functional dependency of I_3 on L_1 and C_2, one needs to do seven measurements of current phasor I_3 for seven different sets of values (L_1, C_2). Suppose that the measurements are taken and let Table 3.1 summarize the numerical values assigned to L_1 and C_2 and the corresponding measurements of I_3. For this case, the general functional dependency can be written as

$$I_3(L_1, C_2) = \frac{\alpha_0 + \alpha_1 L_1 j\omega_0 + \alpha_2/(C_2 j\omega_0) + \alpha_3 L_1/C_2}{\beta_0 + \beta_1 L_1 j\omega_0 + \beta_2/(C_2 j\omega_0) + L_1/C_2}, \tag{3.13}$$

where the complex constants $\alpha_0, \alpha_1, \alpha_2, \alpha_3, \beta_0, \beta_1, \beta_2$ can be determined by solving the set of seven measurement equations.

Substituting the numerical values given in Table 3.1 into the measurement equations and solving for the unknown complex constants yields

$$\begin{aligned} \alpha_0 &= -1502 - 2772j, & \alpha_1 &= 173 + 74j, \\ \alpha_2 &= 106 + 151j, & \alpha_3 &= 0, \\ \beta_0 &= -481 - 316j, & \beta_1 &= 13 + 13j, \\ \beta_2 &= 30 + 15j. & & \end{aligned} \tag{3.14}$$

Table 3.1 Numerical values of the measurements for the AC circuit example

Exp. no.	L_1 (m H)	C_2 (μ F)	I_3 (A)
1	13	10	3.3−2.9i
2	25	20	2.7−3.2i
3	32	23	2.3−3.4i
4	45	29	1.4−3.6i
5	54	33	.7−3.5i
6	68	40	−.5−2.9i
7	90	47	−1.4−1.3i

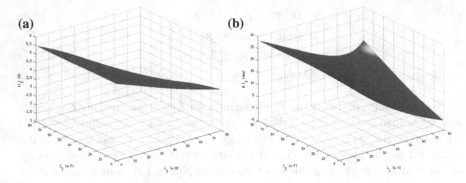

Fig. 3.5 $I_3(j\omega_0)$ versus L_1 and C_2. **a** $|I_3(j\omega_0)|$ vs. L_1 and C_2, **b** $\angle I_3(j\omega_0)$ vs. L_1 and C_2

Thus, the functional dependency of I_3 on L_1 and C_2 will be

$$I_3(L_1, C_2) = \frac{(-1502 - 2772j) + (173 + 74j)L_1 j\omega_0 + (106 + 151j)/(C_2 j\omega_0)}{(-481 - 316j) + (13 + 13j)L_1 j\omega_0 + (30 + 15j)/(C_2 j\omega_0) + L_1/C_2}. \tag{3.15}$$

Figure 3.5a, b, show the plots of the magnitude and the phase of I_3 as a function of the design elements L_1 and C_2, as obtained in (3.15). Applying constraints (3.9) and (3.10) on I_3, one may obtain the region in the L_1–C_2 plane, shown in black color in Fig. 3.6, where these constraints are satisfied.

(b) I_9 versus L_1 and C_2

Following the same procedure, one may obtain the dependency of I_9 on L_1 and C_2. Plots of the magnitude and the phase of I_9 as a function of L_1 and C_2 are shown

Fig. 3.6 Region (in *black color*) where (3.9) and (3.10) are satisfied

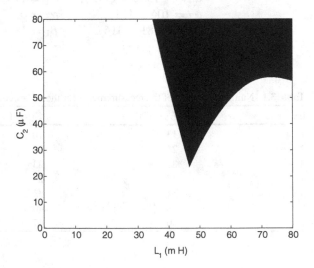

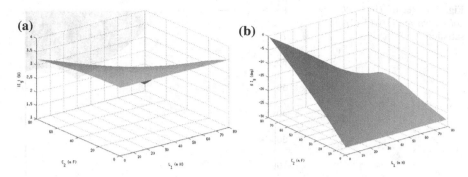

Fig. 3.7 $I_9(j\omega_0)$ versus L_1 and C_2. **a** $|I_9(j\omega_0)|$ vs. L_1 and C_2, **b** $\angle I_9(j\omega_0)$ vs. L_1 and C_2

in Fig. 3.7a, b, respectively. Applying constraints (3.11) and (3.12) on I_9, one can find the region in the L_1–C_2 plane, shown in black color in Fig. 3.8, where these constraints are valid.

In order to satisfy the constraints given in (3.9)–(3.12), simultaneously, one has to intersect the regions shown in Figs. 3.6 and 3.8. Figure 3.9 shows the region in the L_1–C_2 plane where constraints (3.9)–(3.12) are simultaneously satisfied.

Fig. 3.8 Region (in *black color*) where (3.11) and (3.12) are satisfied

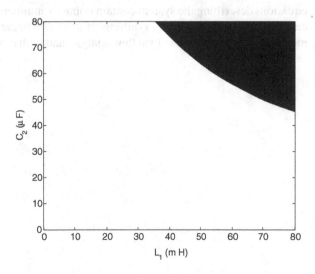

Fig. 3.9 Region (in *black color*) where (3.9)–(3.12) are satisfied

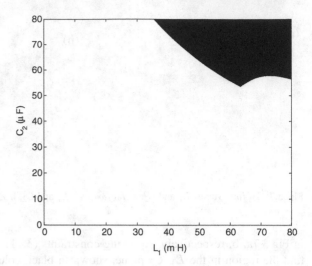

3.4 Notes and References

In this chapter we extended the measurement based approach developed in Chap. 2 for DC circuits, to linear AC circuits. The main difference here is that the linear equations describing the system contain complex numbers. All the results of Chap. 2 carry over to the analysis and synthesis of unknown linear AC circuits. These results may be useful in model free load flow analysis and fault monitoring in power systems.

Chapter 4
Application to Mechanical Systems

This chapter presents the application of the measurement based approach to linear mechanical systems, civil engineering truss structures and linear hydraulic networks.

4.1 Mass-Spring Systems

In this section we consider synthesis problems where in an unknown mass-spring system the displacements of the masses are to be controlled by adjusting the spring stiffness constants. Consider the unknown linear mass-spring system shown in Fig. 4.1.

Suppose that we want to control the displacement of the ith mass, denoted by x_i, by adjusting the spring stiffness k_j at an arbitrary location. Assume that the spring k_j is composed of piezoelectric materials such that its stiffness can be controlled by applying an electrical field. The displacements can be measured using a variety of sensors such as potentiometers or Linear Variable Differential Transformers (LVDTs). In this problem the system of governing linear equations can be constructed in the form:

$$\underbrace{\begin{bmatrix} k_1 + k_2 & -k_2 & 0 & \cdots & 0 & 0 \\ -k_2 & k_2 + k_3 & -k_3 & \cdots & 0 & 0 \\ 0 & -k_3 & k_3 + k_4 & \cdots & 0 & 0 \\ \vdots & \vdots & \vdots & & \vdots & \vdots \\ 0 & 0 & 0 & \cdots & -k_{n-1} & k_n \end{bmatrix}}_{A(p)} \underbrace{\begin{bmatrix} x_1 \\ x_2 \\ x_3 \\ \vdots \\ x_n \end{bmatrix}}_{x} = \underbrace{\begin{bmatrix} F_1 \\ F_2 \\ F_3 \\ \vdots \\ F_n \end{bmatrix}}_{b(q)}, \qquad (4.1)$$

where $p = [k_1, k_2, \ldots, k_n]^T$, x represents the vector of unknown displacements and $q = [F_1, F_2, \ldots, F_n]^T$ is the vector of external forces. Applying the Cramer's rule to (4.1) to calculate x_i gives

S. P. Bhattacharyya et al., *Linear Systems*, SpringerBriefs in Applied Sciences and Technology, DOI: 10.1007/978-81-322-1641-4_4, © The Author(s) 2014

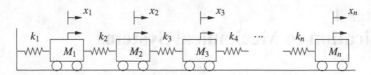

Fig. 4.1 An unknown general mass-spring system

$$x_i(\mathbf{p}, \mathbf{q}) = \frac{|\mathbf{B}_i(\mathbf{p}, \mathbf{q})|}{|\mathbf{A}(\mathbf{p})|}, \quad i = 1, 2, \ldots, n, \tag{4.2}$$

where $\mathbf{B}_i(\mathbf{p}, \mathbf{q})$ is the matrix obtained by replacing the ith column of $\mathbf{A}(\mathbf{p})$ by $\mathbf{b}(\mathbf{q})$. We can state the following theorem.

Theorem 4.1 *In a linear mass-spring system, with fixed external forces and fixed spring constants except k_j, the functional dependency of any mass displacement x_i on any spring stiffness k_j can be determined by 3 measurements of the displacement x_i obtained for 3 different values of k_j.*

Proof Note that the matrices $\mathbf{B}_i(\mathbf{p}, \mathbf{q})$ and $\mathbf{A}(\mathbf{p})$, in (4.2), are both of rank 1 with respect to k_j. According to Lemma 1.1, the functional dependency of x_i on k_j can be expressed as

$$x_i(k_j) = \frac{\tilde{\alpha}_0 + \tilde{\alpha}_1 k_j}{\tilde{\beta}_0 + \tilde{\beta}_1 k_j}, \tag{4.3}$$

where $\tilde{\alpha}_0, \tilde{\alpha}_1, \tilde{\beta}_0, \tilde{\beta}_1$ are constants. If $\tilde{\beta}_0 = \tilde{\beta}_1 = 0$, then $x_i \to \infty$, for any value of the spring stiffness k_j, which is physically impossible. Hence, we rule out this case. Assuming that $\tilde{\beta}_1 \neq 0$, one can divide the numerator and denominator of (4.3) by $\tilde{\beta}_1$ and obtain

$$x_i(k_j) = \frac{\alpha_0 + \alpha_1 k_j}{\beta_0 + k_j}, \tag{4.4}$$

where $\alpha_0, \alpha_1, \beta_0$ are constants. In order to determine $\alpha_0, \alpha_1, \beta_0$ one conducts 3 experiments by setting 3 different values to the spring stiffness k_j, namely k_{j1}, k_{j2}, k_{j3} and measuring the corresponding displacements x_i, namely x_{i1}, x_{i2}, x_{i3}. The following set of measurement equations can then be formed:

$$\underbrace{\begin{bmatrix} 1 & k_{j1} & -x_{i1} \\ 1 & k_{j2} & -x_{i2} \\ 1 & k_{j3} & -x_{i3} \end{bmatrix}}_{\mathbf{M}} \underbrace{\begin{bmatrix} \alpha_0 \\ \alpha_1 \\ \beta_0 \end{bmatrix}}_{\mathbf{u}} = \underbrace{\begin{bmatrix} x_{i1} k_{j1} \\ x_{i2} k_{j2} \\ x_{i3} k_{j3} \end{bmatrix}}_{\mathbf{m}}. \tag{4.5}$$

The set of measurement equations (4.5) can be uniquely solved for the unknown constants $\alpha_0, \alpha_1, \beta_0$ provided that $|\mathbf{M}| \neq 0$. If $|\mathbf{M}| = 0$, the last column of the matrix $\mathbf{M}$ can be expressed as a linear combination of the first two columns because by assigning different values to the spring stiffness k_j, the first two columns of $\mathbf{M}$ become linearly independent. In this case, the functional dependency of x_i on k_j can be will be

$$x_i(k_j) = \alpha_0 + \alpha_1 k_j, \tag{4.6}$$

where α_0, α_1 are constants that can be determined from any two of the experiments conducted earlier. The functional dependency (4.6) corresponds to the case where $\tilde{\beta}_1 = 0$ in (4.3), and the numerator and denominator of (4.3) are divided by $\tilde{\beta}_0$.   □

Remark 1 Suppose that the design parameters are the external forces applied to the system. In such a case, the displacement x_i can be expressed as

$$x_i(F_1, F_2, \ldots, F_n) = \beta_1 F_1 + \beta_2 F_2 + \cdots + \beta_n F_n, \tag{4.7}$$

and the constants $\beta_1, \beta_2, \ldots, \beta_n$ can be determined by applying n different sets of vector forces $(F_1, F_2, \ldots, F_n)$ to the system and measuring the corresponding displacements x_i. This is the well-known Superposition Principle in mechanical systems. In addition, if the external forces vary in the intervals $F_j^- \le F_j \le F_j^+$, $j = 1, 2, \ldots, n$, then the displacement x_i will vary in a convex hull whose vertices can be computed using the vertices (F_j^-, F_j^+), $j = 1, 2, \ldots, n$.

4.2 Truss Structures

Here we consider the class of truss structures and want to control the displacements of some set of joints by adjusting the stiffness of some set of design elements. Suppose that Fig. 4.2 represents an unknown general truss structure.

Fig. 4.2 An unknown general truss structure

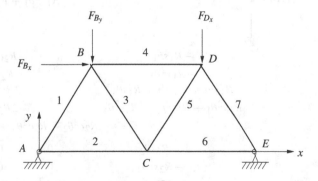

The element-wise stiffness matrix K_k, associated with the element k, can be constructed as

$$K_k = \frac{E_k A_k}{L_k} \begin{bmatrix} \cos^2 \theta_k & \frac{1}{2} \sin 2\theta_k & -\cos^2 \theta_k & -\frac{1}{2} \sin 2\theta_k \\ \frac{1}{2} \sin 2\theta_k & \sin^2 \theta_k & -\frac{1}{2} \sin 2\theta_k & -\sin^2 \theta_k \\ -\cos^2 \theta_k & -\frac{1}{2} \sin 2\theta_k & \cos^2 \theta_k & \frac{1}{2} \sin 2\theta_k \\ -\frac{1}{2} \sin 2\theta_k & -\sin^2 \theta_k & \frac{1}{2} \sin 2\theta_k & \sin^2 \theta_k \end{bmatrix}, \quad (4.8)$$

where E_k denotes the modulus of elasticity, A_k is the cross section area, L_k is the length of the element and θ_k is the angle of the element. For the sake of simplicity let us define $R_k := E_k A_k / L_k$. The global stiffness matrix $\mathbf{A}(\mathbf{p})$ in (4.9) can then formed from element-wise stiffness matrices. Let s and c denote $\sin(.)$ and $\cos(.)$ functions, respectively. Then, the governing linear equations, for the truss structure shown in Fig. 4.2, can be written in the following matrix form:

$$\underbrace{\begin{bmatrix} \mathbf{A}_{11}(\mathbf{p}) & \mathbf{A}_{12}(\mathbf{p}) \\ \mathbf{A}_{12}^T(\mathbf{p}) & \mathbf{A}_{22}(\mathbf{p}) \end{bmatrix}}_{\mathbf{A}(\mathbf{p})} \underbrace{\begin{bmatrix} \delta_{Ax} \\ \delta_{Ay} \\ \vdots \\ \delta_{Ex} \\ \delta_{Ey} \end{bmatrix}}_{\mathbf{x}} = \underbrace{\begin{bmatrix} F_{Ax} \\ F_{Ay} \\ \vdots \\ F_{Ex} \\ F_{Ey} \end{bmatrix}}_{\mathbf{b}(\mathbf{q})}, \quad (4.9)$$

where $\mathbf{A}(\mathbf{p})$ is the global stiffness matrix,

$$\mathbf{A}_{11}(\mathbf{p}) = \begin{bmatrix} R_1 c^2\theta_1 + R_2 c^2\theta_2 & R_1 s\theta_1 + R_2 s\theta_2 & -R_1 c^2\theta_1 \\ & R_1 s^2\theta_1 + R_2 s^2\theta_2 & -R_1 s\theta_1 \\ & & R_1 c^2\theta_1 + R_3 c^2\theta_3 + R_4 c^2\theta_4 \end{bmatrix},$$

$$\begin{bmatrix} -R_1 s\theta_1 & -R_2 c^2\theta_2 \\ -R_1 s^2\theta_1 & -R_2 s\theta_2 \\ R_1 s\theta_1 + R_3 s\theta_3 + R_4 s\theta_4 & -2R_3 c^2\theta_3 \\ R_1 s^2\theta_1 + R_3 s^2\theta_3 + R_4 s^2\theta_4 & -2R_3 s\theta_3 \\ & R_2 c^2\theta_2 + R_3 c^2\theta_3 + R_5 c^2\theta_5 + R_6 c^2\theta_6 \end{bmatrix},$$

$$\mathbf{A}_{12}(\mathbf{p}) = \begin{bmatrix} -R_2 s\theta_2 & 0 & 0 & 0 & 0 \\ -R_2 s^2\theta_2 & 0 & 0 & 0 & 0 \\ -2R_3 s\theta_3 & -R_4 c^2\theta_4 & -R_4 s\theta_4 & 0 & 0 \\ -2R_3 s^2\theta_3 & -R_4 s\theta_4 & -R_4 s^2\theta_4 & 0 & 0 \\ R_2 s\theta_2 + R_3 s\theta_3 + R_5 s\theta_5 + R_6 s\theta_6 & -R_5 c^2\theta_5 & -R_5 s\theta_5 & -R_6 c^2\theta_6 & -R_6 s\theta_6 \end{bmatrix},$$

$$
\mathbf{A}_{22}(\mathbf{p}) =
\begin{bmatrix}
R_2 s^2 \theta_2 + R_3 s^2 \theta_3 + R_5 s^2 \theta_5 + R_6 s^2 \theta_6 & -R_5 s \theta_5 \\
& R_4 c^2 \theta_4 + R_5 c^2 \theta_5 + R_7 c^2 \theta_7 \\
\\
-R_5 s^2 \theta_5 & -R_6 s \theta_6 & -R_6 s^2 \theta_6 \\
R_4 s \theta_4 + R_5 s \theta_5 + R_7 s \theta_7 & -R_7 c^2 \theta_7 & -R_7 s \theta_7 \\
R_4 s^2 \theta_4 + R_5 s^2 \theta_5 + R_7 s^2 \theta_7 & -R_7 s \theta_7 & -R_7 s \theta_7 \\
& R_6 c^2 \theta_6 + R_7 c^2 \theta_7 & R_6 s \theta_6 + R_7 s \theta_7 \\
& & R_6 s^2 \theta_6 + R_7 s^2 \theta_7
\end{bmatrix},
$$

and $\mathbf{p}$ denotes the vector of elements parameters, $\mathbf{x}$ is the vector of unknown joints displacements (in x and y directions) and $\mathbf{q}$ represents the vector of external forces applied to the truss structure. Similar to the previous section, if one applies the Cramer's rule to (4.9) to calculate the ith component of x, denoted by x_i, then

$$
x_i(\mathbf{p}, \mathbf{q}) = \frac{|\mathbf{B}_i(\mathbf{p}, \mathbf{q})|}{|\mathbf{A}(\mathbf{p})|}, \quad i = 1, 2, \ldots, n, \tag{4.10}
$$

where $\mathbf{B}_i(\mathbf{p}, \mathbf{q})$ is the matrix $\mathbf{A}(\mathbf{p})$ with the ith column replaced by $\mathbf{b}(\mathbf{q})$.

Assuming that the design elements are composed of piezoelectric materials, one can control their cross section areas, and thus their stiffness constants, by applying an electrical field. Suppose that in this problem, the design parameters are the cross section areas of some set of design elements. We have the following theorem.

Theorem 4.2 *In a linear truss structure, with fixed external forces and fixed element stiffness constants except element cross section area A_j, the functional dependency of a given joint displacement δ_i, at a given direction, on A_j can be determined by 3 measurements of the joint displacement δ_i, in the respective direction, obtained for 3 different values of A_j.*

Proof The proof is similar to the previous problem (Sect. 4.1). Recalling the procedure of assembling element-wise stiffness matrices into the global stiffness matrix $\mathbf{A}(\mathbf{p})$, it can be concluded that the matrices $\mathbf{B}_i(\mathbf{p}, \mathbf{q})$ and $\mathbf{A}(\mathbf{p})$, in (4.10), are both of rank 1 with respect to A_j. According to Lemma 1.1, the functional dependency of δ_i on A_j can be expressed as

$$
\delta_i(A_j) = \frac{\alpha_0 + \alpha_1 A_j}{\beta_0 + A_j}, \tag{4.11}
$$

where $\alpha_0, \alpha_1, \beta_0$ are constants that can be determined by conducting 3 experiments. □

4.3 Hydraulic Networks

In this section we extend our measurement based approach to linear hydraulic networks. Suppose that, in a hydraulic network, all the flows are in the laminar state resulting in the governing steady state equations to be linear. Here the objective is to control the flow rates passing through some set of pipes.

In a laminar flow, the pressure drop occurring in a pipe can be obtained from,

$$\Delta P = \frac{8\mu L Q}{\pi r^4},\tag{4.12}$$

where μ is the dynamic viscosity of the fluid, L is the length of the pipe, Q is the volume flow rate and r is the inner radius of the pipe. Let us rewrite (4.12) as

$$\Delta P = RQ,\tag{4.13}$$

where

$$R = \frac{8\mu L}{\pi r^4},\tag{4.14}$$

is called the pipe resistance constant which is a function of the mechanical properties (length L and radius r) of the pipe.

To illustrate the approach, we begin by considering an unknown general hydraulic network as shown in Fig. 4.3.

Similar to linear circuits, applying Kirchhoff's laws to a linear hydraulic network (Fig. 4.3) gives the set of governing linear equations shown below in matrix form:

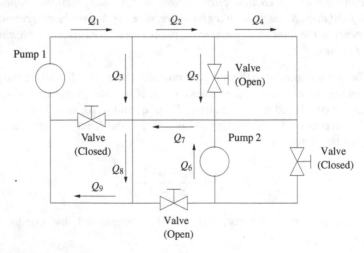

Fig. 4.3 An unknown general hydraulic network

$$\underbrace{\begin{bmatrix} 1 & -1 & -1 & 0 & 0 & 0 & 0 & 0 \\ 0 & 1 & 0 & -1 & -1 & 0 & 0 & 0 \\ 0 & 0 & 0 & 1 & 1 & 1 & -1 & 0 \\ 0 & 0 & 1 & 0 & 0 & 0 & 1 & -1 \\ R_1 & 0 & R_3 & 0 & 0 & 0 & 0 & R_8 \\ 0 & -R_2 & R_3 & 0 & -R_5 & R_6 & 0 & R_8 \\ 0 & 0 & 0 & -R_4 & R_5 & 0 & 0 & 0 \\ 0 & -R_2 & R_3 & -R_4 & 0 & 0 & -R_7 & 0 \end{bmatrix}}_{\mathbf{A(p)}} \underbrace{\begin{bmatrix} Q_1 \\ Q_2 \\ Q_3 \\ Q_4 \\ Q_5 \\ Q_6 \\ Q_7 \\ Q_8 \end{bmatrix}}_{\mathbf{x}} = \underbrace{\begin{bmatrix} 0 \\ 0 \\ 0 \\ 0 \\ P_1 \\ P_2 \\ 0 \\ 0 \end{bmatrix}}_{\mathbf{b(q)}}, \qquad (4.15)$$

where $\mathbf{p} = [R_1, R_2, \ldots, R_8]$ is the vector of the pipe resistances (R_i, $i = 1, 2, \ldots, 8$ is the resistance of the set of pipes through which Q_i, $i = 1, 2, \ldots, 8$ flows), $\mathbf{x}$ denotes the vector of unknown flow rates and $\mathbf{q}$ represents the vector of input parameters including the pump pressures. The flow rate Q_i can be calculated from (4.15) using the Cramer's rule,

$$Q_i = x_i(\mathbf{p}, \mathbf{q}) = \frac{|\mathbf{B}_i(\mathbf{p}, \mathbf{q})|}{|\mathbf{A(p)}|}, \quad i = 1, 2, \ldots, n, \qquad (4.16)$$

where $\mathbf{B}_i(\mathbf{p}, \mathbf{q})$ is the matrix $\mathbf{A(p)}$ with the ith column replaced by $\mathbf{b(q)}$.

Observation 4.1 Upon an application of Kirchhoff's laws, each pipe resistance R_j appears in only one column of the characteristic matrix $\mathbf{A(p)}$.

In hydraulic networks, it is common to carry out the measurements using the pressure-driven test. In the following design cases, assume that the set of design pipes are composed of piezoelectric materials, and thus the radii of these pipes can be controlled upon applying an electrical field. In case the radius can not be controlled by applying electrical fields, different pilot pipe sections have to be used for the experiments.

4.3.1 Flow Rate Control Using a Single Pipe Resistance

Assume that the design parameter is the resistance of one pipe, denoted by R_j, at an arbitrary location of the network. Therefore, we want to control the flow rate at some location of the network, denoted by Q_i, by adjusting the pipe resistance R_j. We can state the following theorem.

Theorem 4.3 *In a linear hydraulic network, with fixed pipe resistances except R_j, the functional dependency of any flow rate Q_i on the pipe resistance R_j can be determined by at most 3 measurements of the flow rate Q_i obtained for 3 different values of R_j.*

Proof Let us consider two cases: (1) $i \neq j$ and (2) $i = j$.

Case 1: $i \neq j$

Based on the Observation 4.1, the matrices $\mathbf{B}_i(\mathbf{p}, \mathbf{q})$ and $\mathbf{A}(\mathbf{p})$, in (4.16), are both of rank 1 with respect to R_j. Therefore, according on Lemma 1.1, the functional dependency of Q_i on R_j can be expressed as

$$Q_i(R_j) = \frac{\tilde{\alpha}_0 + \tilde{\alpha}_1 R_j}{\tilde{\beta}_0 + \tilde{\beta}_1 R_j}, \tag{4.17}$$

where $\tilde{\alpha}_0, \tilde{\alpha}_1, \tilde{\beta}_0, \tilde{\beta}_1$ are constants. Assume that $\tilde{\beta}_1 \neq 0$; hence, one can divide the numerator and denominator of (4.17) by $\tilde{\beta}_1$ and obtain

$$Q_i(R_j) = \frac{\alpha_0 + \alpha_1 R_j}{\beta_0 + R_j}, \tag{4.18}$$

where $\alpha_0, \alpha_1, \beta_0$ are constants that can be determined by conducting 3 experiments. The measurement equations can be written as

$$\underbrace{\begin{bmatrix} 1 & R_{j1} & -Q_{i1} \\ 1 & R_{j2} & -Q_{i2} \\ 1 & R_{j3} & -Q_{i3} \end{bmatrix}}_{\mathbf{M}} \underbrace{\begin{bmatrix} \alpha_0 \\ \alpha_1 \\ \beta_0 \end{bmatrix}}_{\mathbf{u}} = \underbrace{\begin{bmatrix} Q_{i1} R_{j1} \\ Q_{i2} R_{j2} \\ Q_{i3} R_{j3} \end{bmatrix}}_{\mathbf{m}}, \tag{4.19}$$

which can be uniquely solved for the unknown constants $\alpha_0, \alpha_1, \beta_0$, provided $|\mathbf{M}| \neq 0$. For the situations where $|\mathbf{M}| = 0$, corresponding to $\tilde{\beta}_1 = 0$ in (4.17), one may use a similar strategy presented in Sect. 2.2.1 to derive the corresponding functional dependency. In this case,

$$Q_i(R_j) = \alpha_0 + \alpha_1 R_j. \tag{4.20}$$

Case 2: $i = j$

Recalling (4.16) and based on the Observation 4.1, the matrix $\mathbf{B}_i(\mathbf{p}, \mathbf{q})$ is of rank 0 with respect to R_i; however, the matrix $\mathbf{A}(\mathbf{p})$ is of rank 1 with respect to R_j. According to Lemma 1.1, the functional dependency of Q_i on R_i will be

$$Q_i(R_i) = \frac{\tilde{\alpha}_0}{\tilde{\beta}_0 + \tilde{\beta}_1 R_i}, \tag{4.21}$$

where $\tilde{\alpha}_0, \tilde{\beta}_0, \tilde{\beta}_1$ are constants. Assuming that $\tilde{\beta}_1 \neq 0$, and dividing the numerator and denominator of (4.21) by $\tilde{\beta}_1$ results in

$$Q_i(R_i) = \frac{\alpha_0}{\beta_0 + R_i}, \tag{4.22}$$

where α_0, β_0 are constants that can be determined by conducting 2 experiments. The following set of measurement equations can then be formed

$$\underbrace{\begin{bmatrix} 1 & -Q_{i1} \\ 1 & -Q_{i2} \end{bmatrix}}_{\mathbf{M}} \underbrace{\begin{bmatrix} \alpha_0 \\ \beta_0 \end{bmatrix}}_{\mathbf{u}} = \underbrace{\begin{bmatrix} Q_{i1}R_{i1} \\ Q_{i2}R_{i2} \end{bmatrix}}_{\mathbf{m}}, \tag{4.23}$$

and uniquely solved for the unknown constants α_0, β_0, if and only if $|\mathbf{M}| \neq 0$. If $|\mathbf{M}| = 0$ in (4.23), it can be concluded that Q_i is a constant,

$$Q_i(R_i) = \alpha_0. \tag{4.24}$$

This case corresponds to $\tilde{\beta}_1 = 0$ in (4.21). □

4.3.2 Flow Rate Control Using Two Pipe Resistances

Suppose that the design parameters are any two pipe resistances, denoted by R_j and R_k, at arbitrary locations of the network, and the flow rate Q_i, at some location of the network, is to be controlled by adjusting these two pipe resistances.

Theorem 4.4 *In a linear hydraulic network, with fixed pipe resistances except R_j and R_k, the functional dependency of any flow rate Q_i on the pipe resistances R_j and R_k can be determined by at most 7 measurements of the flow rate Q_i obtained for 7 different sets of values of (R_j, R_k).*

Proof Again, let us consider two cases: (1) $i \neq j, k$ and (2) $i = j$ or $i = k$.

Case 1: $i \neq j, k$
In this case, based on the Observation 4.1, the matrices $\mathbf{B}_i(\mathbf{p}, \mathbf{q})$ and $\mathbf{A}(\mathbf{p})$, in (4.16), are both of rank 1 with respect to R_j and R_k. According to Lemma 1.2, the functional dependency of Q_i on R_j and R_k becomes

$$Q_i(R_j, R_k) = \frac{\tilde{\alpha}_0 + \tilde{\alpha}_1 R_j + \tilde{\alpha}_2 R_k + \tilde{\alpha}_3 R_j R_k}{\tilde{\beta}_0 + \tilde{\beta}_1 R_j + \tilde{\beta}_2 R_k + \tilde{\beta}_3 R_j R_k}, \tag{4.25}$$

where $\tilde{\alpha}_0$, $\tilde{\alpha}_1$, $\tilde{\alpha}_2$, $\tilde{\alpha}_3$, $\tilde{\beta}_0$, $\tilde{\beta}_1$, $\tilde{\beta}_2$, $\tilde{\beta}_3$ are constants. Assuming that $\tilde{\beta}_3 \neq 0$, dividing the numerator and denominator of (4.25) by $\tilde{\beta}_3$ yields

$$Q_i(R_j, R_k) = \frac{\alpha_0 + \alpha_1 R_j + \alpha_2 R_k + \alpha_3 R_j R_k}{\beta_0 + \beta_1 R_j + \beta_2 R_k + R_j R_k}, \tag{4.26}$$

where α_0, α_1, α_2, α_3, β_0, β_1, β_2 are constants. In order to determine these constants, one conducts 7 experiments by setting 7 different sets of valuesto the pipe resistances

(R_j, R_k), and measuring the corresponding flow rates Q_i. The following set of measurement equations can then be obtained

$$\underbrace{\begin{bmatrix} 1 & R_{j1} & R_{k1} & R_{j1}R_{k1} & -Q_{i1} & -Q_{i1}R_{j1} & -Q_{i1}R_{k1} \\ 1 & R_{j2} & R_{k2} & R_{j2}R_{k2} & -Q_{i2} & -Q_{i2}R_{j2} & -Q_{i2}R_{k2} \\ 1 & R_{j3} & R_{k3} & R_{j3}R_{k3} & -Q_{i3} & -Q_{i3}R_{j3} & -Q_{i3}R_{k3} \\ 1 & R_{j4} & R_{k4} & R_{j4}R_{k4} & -Q_{i4} & -Q_{i4}R_{j4} & -Q_{i4}R_{k4} \\ 1 & R_{j5} & R_{k5} & R_{j5}R_{k5} & -Q_{i5} & -Q_{i5}R_{j5} & -Q_{i5}R_{k5} \\ 1 & R_{j6} & R_{k6} & R_{j6}R_{k6} & -Q_{i6} & -Q_{i6}R_{j6} & -Q_{i6}R_{k6} \\ 1 & R_{j7} & R_{k7} & R_{j7}R_{k7} & -Q_{i7} & -Q_{i7}R_{j7} & -Q_{i7}R_{k7} \end{bmatrix}}_{\mathbf{M}} \underbrace{\begin{bmatrix} \alpha_0 \\ \alpha_1 \\ \alpha_2 \\ \alpha_3 \\ \beta_0 \\ \beta_1 \\ \beta_2 \end{bmatrix}}_{\mathbf{u}} = \underbrace{\begin{bmatrix} Q_{i1}R_{j1}R_{k1} \\ Q_{i2}R_{j2}R_{k2} \\ Q_{i3}R_{j3}R_{k3} \\ Q_{i4}R_{j4}R_{k4} \\ Q_{i5}R_{j5}R_{k5} \\ Q_{i6}R_{j6}R_{k6} \\ Q_{i7}R_{j7}R_{k7} \end{bmatrix}}_{\mathbf{m}}.$$

$$(4.27)$$

This set of equations can be uniquely solved for the constants $\alpha_0, \alpha_1, \alpha_2, \alpha_3$, $\beta_0, \beta_1, \beta_2$, provided $|\mathbf{M}| \neq 0$. If $|\mathbf{M}| = 0$, one can follow a similar procedure presented in Sect. 2.2.2 to develop the corresponding functional dependency of Q_i on R_j and R_k.

Case 2: $i = j$ or $i = k$

Suppose that $i = j$ and recall (4.16). Based on the Observation 4.1, in this case, the matrix $\mathbf{A}(\mathbf{p})$ is of rank 1 with respect to R_i and R_k; however, the matrix $\mathbf{B}_i(\mathbf{p}, \mathbf{q})$ is of rank 0 with respect to R_i and is of rank 1 with respect to R_k. According to Lemma 1.2 and these rank conditions, the functional dependency of Q_i on R_i and R_k will be

$$Q_i(R_i, R_k) = \frac{\tilde{\alpha}_0 + \tilde{\alpha}_1 R_k}{\tilde{\beta}_0 + \tilde{\beta}_1 R_i + \tilde{\beta}_2 R_k + \tilde{\beta}_3 R_i R_k}, \qquad (4.28)$$

where $\tilde{\alpha}_0, \tilde{\alpha}_1, \tilde{\beta}_0, \tilde{\beta}_1, \tilde{\beta}_2, \tilde{\beta}_3$ are constants. Assuming that $\tilde{\beta}_3 \neq 0$, one can divide the numerator and denominator of (4.28) by $\tilde{\beta}_3$ and obtain

$$Q_i(R_i, R_k) = \frac{\alpha_0 + \alpha_1 R_k}{\beta_0 + \beta_1 R_i + \beta_2 R_k + R_i R_k}, \qquad (4.29)$$

where $\alpha_0, \alpha_1, \beta_0, \beta_1, \beta_2$ are constants that can be determined by conducting 5 experiments, by assigning 5 different sets of values to the pipe resistances (R_i, R_k), measuring the corresponding flow rates Q_i, and solving the obtained set of equations. $\qquad\qquad\qquad\qquad\qquad\qquad\qquad\qquad\qquad\qquad\qquad\qquad\square$

4.4 Illustrative Examples

4.4.1 An Example of Mass-Spring Systems

Consider the three-story building frame shown in Fig. 4.4a (A similar two-story building frame example can be found in [1, p. 362, exercise 5.24]). Suppose that

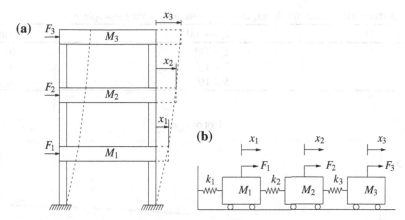

Fig. 4.4 An example of mass-spring systems **a** A three-story building **b** Mass-spring model of the three-story building

the mechanical properties of the building components are unknown and the building is modeled as a mass-spring system shown is Fig. 4.4b with unknown parameters. Assume that the links connecting the first and the second floors are composed of piezoelectric materials and thus their stiffness can be controlled by applying an electrical field.

Suppose that the design objective is to control the displacement of the second floor (mass M_2), denoted by x_2, by adjusting the stiffness constants of the links connecting the first and the second floors (spring constant k_2), to be within the range

$$-0.05 \leq x_2 \leq -0.03 \ (m). \tag{4.30}$$

Hence, we need to find an interval of k_2 values for which (4.30) is satisfied. Based on Theorem 4.1 the functional dependency of x_2 on k_2 can be written as

$$x_2 = \frac{\alpha_0 + \alpha_1 k_2}{\beta_0 + k_2}, \tag{4.31}$$

where α_0, α_1, β_0 are constants that can be determined by conducting 3 experiments, by setting 3 different values to the spring constant k_2, namely k_{21}, k_{22}, k_{23}, measuring the corresponding displacements x_2, namely x_{21}, x_{22}, x_{23}, and then solving the following system of measurement equations

$$\begin{bmatrix} 1 & k_{21} & -x_{21} \\ 1 & k_{22} & -x_{22} \\ 1 & k_{23} & -x_{23} \end{bmatrix} \begin{bmatrix} \alpha_0 \\ \alpha_1 \\ \beta_0 \end{bmatrix} = \begin{bmatrix} k_{21}x_{21} \\ k_{22}x_{22} \\ k_{23}x_{23} \end{bmatrix}. \tag{4.32}$$

Suppose that the measurements are taken and let Table 4.1 show the numerical values for the experiments performed for this example. Substituting these numerical values

Table 4.1 Numerical values for the experiments of the mass-spring example

Exp. No.	k_2 (N/m)	x_2 (m)
1	2×10^5	-0.035
2	3×10^5	-0.030
3	5×10^5	-0.026

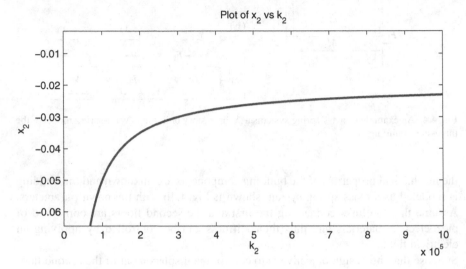

Fig. 4.5 x_2 versus k_2 for the mass-spring example

into (4.32) and solving for the constants yields

$$x_2 = \frac{-3000 - 0.02k_2}{k_2}, \tag{4.33}$$

which is plotted in Fig. 4.5. Applying the design constraint given in (4.30) yields the following range of k_2 values:

$$10^5 \leq k_2 \leq 3 \times 10^5 \ (N/m). \tag{4.34}$$

4.4.2 An Example of Truss Structures

Consider the truss structure shown in Fig. 4.6 (see [2, p. 196, example 4.6.1]) with unknown parameters.

Assuming that the link AC is composed of piezoelectric materials, one can control the cross section area of this link by applying an electrical field. Suppose that we

Fig. 4.6 An example of truss
structures

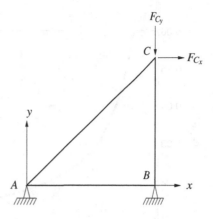

want to control the deflection of the joint "C" in the x-direction, denoted by δ_{Cx}, to
be within the range

$$0 \le \delta_{Cx} \le 0.02 \ (m), \tag{4.35}$$

by adjusting the cross section area of the link AC, A_{AC}.

Based on Theorem 4.2, the functional dependency of δ_{Cx} on A_{AC} can be written as

$$\delta_{Cx} = \frac{\alpha_0 + \alpha_1 A_{AC}}{\beta_0 + A_{AC}}, \tag{4.36}$$

where $\alpha_0, \alpha_1, \beta_0$ are constants that can be determined by conducting 3 experiments
and solving the following system of equations

$$\begin{bmatrix} 1 & A_{AC1} & -\delta_{Cx1} \\ 1 & A_{AC2} & -\delta_{Cx2} \\ 1 & A_{AC3} & -\delta_{Cx3} \end{bmatrix} \begin{bmatrix} \alpha_0 \\ \alpha_1 \\ \beta_0 \end{bmatrix} = \begin{bmatrix} A_{AC1}\delta_{Cx1} \\ A_{AC2}\delta_{Cx2} \\ A_{AC3}\delta_{Cx3} \end{bmatrix}. \tag{4.37}$$

Let Table 4.2 summarize the numerical values for the measurements taken for this
example. Substituting these numerical values into (4.37) and solving for the unknown
constants yields

$$\delta_{Cx} = \frac{1.1 \times 10^{-6} + 6.67 \times 10^{-3} A_{AC}}{A_{AC}}, \tag{4.38}$$

Table 4.2 Numerical values for the experiments of the truss structure example

Exp. No.	A_{AC} (m^2)	δ_{Cx} (m)
1	100×10^{-6}	0.018
2	150×10^{-6}	0.014
3	200×10^{-6}	0.012

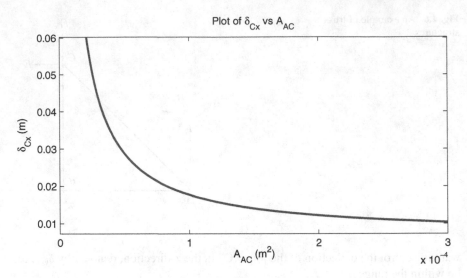

Fig. 4.7 δ_{Cx} versus A_{AC} for the truss structure example

which is plotted in Fig. 4.7. Applying the design constraint, given in (4.35), on δ_{Cx} yields

$$A_{AC} \geq 0.83 \times 10^{-4} \ (m^2). \tag{4.39}$$

4.4.3 An Example of Hydraulic Networks

Consider the unknown hydraulic network shown in Fig. 4.8 and suppose that the flow is laminar, which results in the governing steady state equations to be linear.

Assume that the objective is to control the flow rates Q_8 and Q_{12} (as shown in Fig. 4.8) to stay within the following ranges:

$$0.045 \leq Q_8 \leq 0.055 \ (m^3/s), \tag{4.40}$$

$$0.01 \leq Q_{12} \leq 0.03 \ (m^3/s), \tag{4.41}$$

by adjusting the radii of the pipes numbered 2 and 9, denoted by r_2 and r_9, respectively. Assume that these two pipes are made of piezoelectric materials and their radii can then be adjusted by applying an electrical field. In case the radius can not be controlled by applying electrical fields, seven different pilot pipe sections (for this problem) have to be used for the experiments. Therefore, the design objective is to find regions in the r_2-r_9 plane for which the desired flow rates in (4.40) and (4.41) are met.

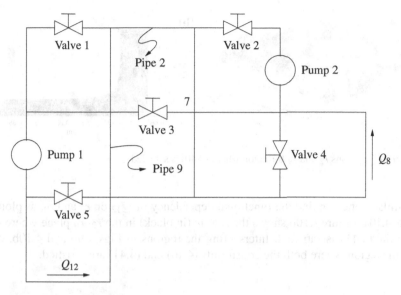

Fig. 4.8 An unknown hydraulic network

Table 4.3 Numerical values for the experiments of the hydraulic network example

Exp. No.	r_2 (m)	R_2 ($Pa.s/m^3$)	r_9 (m)	R_9 ($Pa.s/m^3$)	Q_8 (m^3/s)
1	0.05	408	0.05	408	0.038
2	0.07	107	0.08	62	0.043
3	0.09	39	0.11	17	0.049
4	0.1	26	0.13	9	0.051
5	0.12	12	0.15	5	0.054
6	0.14	6	0.17	3	0.055
7	0.17	3	0.2	1.6	0.056

Based on the results obtained in Sect. 4.3.2, one can determine the functional dependency of any flow rate on any two pipe resistances by at most 7 measurements. Suppose that the measurements are done and let Table 4.3 show the numerical values of the measurements taken to find the functional dependency of Q_8 on r_2 and r_9. Substituting these values into (4.27) and solving for the unknown constants yields the following functional dependency (also recall (4.14))

$$Q_8(r_2, r_9) = \frac{8.7 \times 10^7 + \frac{1600}{r_2^4} + \frac{3500}{r_9^4} + \frac{0.034}{r_2^4 r_9^4}}{1.5 \times 10^9 + \frac{48000}{r_2^4} + \frac{75000}{r_9^4} + \frac{1}{r_2^4 r_9^4}}, \tag{4.42}$$

which is plotted in Fig. 4.9a. Applying the constraint (4.40) to (4.42) yields the region shown (in black) in Fig. 4.9b in the r_2–r_9 plane.

(a) 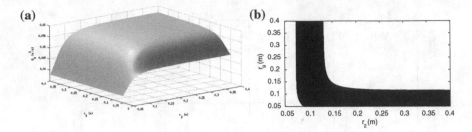 **(b)**

Fig. 4.9 a Q_8 versus r_2 and r_9 **b** Region where (4.40) is satisfied

Similarly one can find the functional dependency of Q_{12} on r_2 and r_9, as plotted in Fig. 4.10a. Figure 4.10b shows the region (in black) in the r_2–r_9 plane where the constraint (4.41) is satisfied. Intersecting the regions in Figs. 4.9b and 4.10b, one finds the region where both the constraints (4.40) and (4.41) are satisfied.

(a) 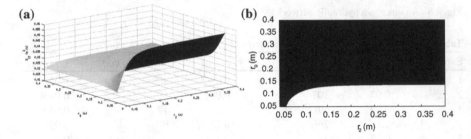 **(b)**

Fig. 4.10 a Q_{12} versus r_2 and r_9 **b** Region where (4.41) is satisfied

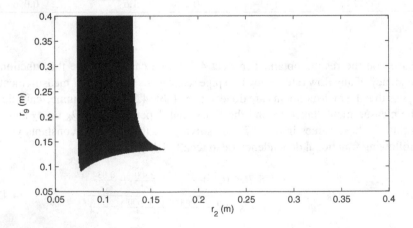

Fig. 4.11 Region where (4.40) and (4.41) are satisfied

4.5 Notes and References

In this chapter we have shown that data from a few measurements can be used to design mass-spring systems, hydraulic systems and truss structures, when the fixed parts of the systems are unknown. These simple applications illustrate the power of the measurement based approach which probably has a much broader range of applications.

The results presented in this chapter are taken from [3]. For more information about piezoelectric materials see [4]. Also, [5] provides some information about sensors such as potentiometers and Linear Variable Differential Transformers (LVDTs). An application of finite element methods in the analysis of truss structures can be found in [2]. For more details about the pressure-driven test which is being used to carry out the measurements in hydraulic networks see [6].

References

1. Rao SS (2000) Mechanical vibrations. Addison-Wesley, New York
2. Reddy JN (2006) An introduction to the finite element method. MacGraw-Hill, New York
3. Mohsenizadeh N, Nounou H, Nounou M, Datta A, Bhattacharyya SP (2013) A measurement based approach to mechanical systems. In the Proceedings of 9th Asian Control Conference, Istanbul, Turkey
4. Tichý J, Erhart J, Kittinger E, Prívratská J (2010) Fundamentals of piezoelectric sensorics: mechanical, dielectric, and thermodynamical properties of piezoelectric materials. Springer, Berlin Heidelberg
5. Nyce DS (2003) Linear position sensors: theory and application. John Wiley & Sons, New Jersey
6. Giustolisi O, Kapelan Z, Savic D (2008) Algorithm for automatic detection of topological changes in water distribution networks. J Hydraul Eng 134(4):435–446

Chapter 5
Application to Control Systems

This chapter presents a new measurement based approach to the problem of controller design for Linear Time-Invariant (LTI) systems directly from the frequency response data. The objective is to guarantee stability and a prescribed desired closed loop frequency response meeting the design specifications. The method to be presented can be used to solve controller design problems wherein the mathematical model of the plant is unknown but for which the frequency response data is available.

5.1 Introduction

The problem of designing controllers satisfying performance requirements has many practical important applications. Most of the classical control design techniques require a mathematical model of the plant, such as a transfer function representation, a priori. In practice one usually deals with very complex systems where modeling is not an easy task and in many cases might be even impossible. In fact, if a model is to be represented for a complex system, it will be of higher order; and consequently, even small perturbations in the model parameters may result in qualitative changes in the behavior of the system, such as transitioning from a stable to an unstable state. This phenomenon is known as *fragility*. These observations motivate a new approach whose objective is to determine the controller parameters directly from measurements without requiring a model of the system. The crux of the idea is that the measurements processed correctly can effectively capture the information necessary to find the design variables directly even though the system equations are unknown.

S. P. Bhattacharyya et al., *Linear Systems*, SpringerBriefs in Applied Sciences and Technology, DOI: 10.1007/978-81-322-1641-4_5, © The Author(s) 2014

5.2 Block Diagrams

We begin by considering the general block diagram shown in Fig. 5.1.
Writing the equations of the system, one can form the matrix equation,

$$
\underbrace{\begin{bmatrix}
1 & 0 & 0 & 0 & 0 & 0 & 0 & 0 & 1 & 0 \\
G_1 & -1 & 0 & 0 & 0 & 0 & 0 & 0 & 0 & 0 \\
0 & -1 & 1 & 0 & 0 & 0 & 0 & 1 & 0 & 0 \\
0 & 0 & G_2 & -1 & 0 & 0 & 0 & 0 & 0 & 0 \\
G_3 & 0 & 0 & 0 & 0 & -1 & 0 & 0 & 0 & 0 \\
0 & 0 & 0 & C_1 & -1 & 0 & 0 & 0 & 0 & 0 \\
0 & 0 & 0 & 0 & 1 & 1 & 1 & 0 & 0 & -1 \\
0 & G_4 & 0 & 0 & 0 & 0 & -1 & 0 & 0 & 0 \\
0 & 0 & 0 & 0 & 0 & 0 & 0 & 0 & -1 & C_2 \\
0 & 0 & 0 & 0 & 0 & 0 & 0 & -1 & 0 & C_3
\end{bmatrix}}_{A(p)}
\underbrace{\begin{bmatrix}
u_1 \\ u_2 \\ u_3 \\ u_4 \\ u_5 \\ u_6 \\ u_7 \\ u_8 \\ u_9 \\ y
\end{bmatrix}}_{x}
=
\underbrace{\begin{bmatrix}
r \\ 0 \\ 0 \\ 0 \\ 0 \\ 0 \\ 0 \\ 0 \\ 0 \\ 0
\end{bmatrix}}_{b(q)},
$$

$$(5.1)$$

where $A(p)$ is the characteristic matrix of the system, x denotes the vector of unknown
signals and $q = q_1 = r$ is the input to the system. It can be easily observed that in
(5.1), the characteristic matrix $A(p)$ is of rank 1 with respect to each of the elements,
G_i, $i = 1, 2, 3, 4$ and C_j, $j = 1, 2, 3$ in the block diagram. Therefore, considering
any of these elements as the design parameter, the transfer function between any two
points of the control system shown in Fig. 5.1 can be expressed as a linear rational
function of the design parameter. For instance, suppose that the design parameter
is C_1, which appears in $A(p)$ with rank 1 dependency. Then closed loop transfer
function between r and y, denoted by $H(s)$, can be represented as

$$
H(s) = \frac{\alpha_0(s) + \alpha_1(s)C_1(s)}{\beta_0(s) + C_1(s)}, \tag{5.2}
$$

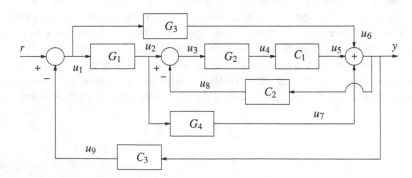

Fig. 5.1 Block diagram of an unknown multivariable system

where $\alpha_0(s)$, $\alpha_1(s)$ and $\beta_0(s)$ are unknown rational functions and are to be determined by performing experiments, as explained in the following section.

5.3 SISO Control Systems

5.3.1 Functional Dependency on a Single Controller

Consider the Single-Input Single-Output (SISO) control system shown in Fig. 5.2. The set of governing equations of this block diagram can be written in the following matrix form:

$$A(p)x = b(q). \tag{5.3}$$

Suppose that the system is unknown and the controller to be designed is denoted by $C(s)$. Recall from the observation presented in the previous section that each element has rank 1 dependency in the system characteristic matrix. We see that we have a situation similar to those derived in the previous chapters. Therefore, the closed loop transfer function can be expressed as

$$H(s) = \frac{\alpha_0(s) + \alpha_1(s)C(s)}{\beta_0(s) + C(s)}, \tag{5.4}$$

where $\alpha_0(s)$, $\alpha_1(s)$ and $\beta_0(s)$ are unknown and $H(s)$ is the transfer function connecting u_e to y_c. Let us rewrite (5.4) in the frequency domain as

$$H(j\omega) = \frac{\alpha_0(j\omega) + \alpha_1(j\omega)C(j\omega)}{\beta_0(j\omega) + C(j\omega)}. \tag{5.5}$$

The unknown coefficients can be determined by embedding three stabilizing controllers, C_1, C_2 and C_3, into the closed loop system and measuring the corresponding closed loop frequency responses, $H_1(j\omega)$, $H_2(j\omega)$ and $H_3(j\omega)$, at a set of frequencies ω_k, $k = 1, 2, \ldots, N$. This can be carried out by exciting the system using a sine

Fig. 5.2 Unknown linear system with controller

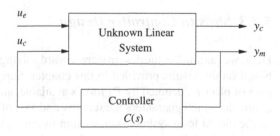

wave at the frequency ω_k (since the system is assumed to be linear, it will respond at the same frequency), and computing the following two quantities:

1. the ratio of the output amplitude to input amplitude, $|H(j\omega_k)|$,
2. the phase shift between output and input, $\angle H(j\omega_k)$,

and finally plotting $|H(j\omega)|$ versus ω and $\angle H(j\omega)$ versus ω, and remembering that $H(j\omega) = |H(j\omega)|e^{\angle H(j\omega)}$. The following system of linear measurement equations can be generated at each frequency $\omega_k, k = 1, 2, \ldots, N$:

$$\underbrace{\begin{bmatrix} 1 & C_1(j\omega_k) & -H_1(j\omega_k) \\ 1 & C_2(j\omega_k) & -H_2(j\omega_k) \\ 1 & C_3(j\omega_k) & -H_3(j\omega_k) \end{bmatrix}}_{\mathbf{M}} \underbrace{\begin{bmatrix} \alpha_0(j\omega_k) \\ \alpha_1(j\omega_k) \\ \beta_0(j\omega_k) \end{bmatrix}}_{\mathbf{u}} = \underbrace{\begin{bmatrix} H_1(j\omega_k)C_1(j\omega_k) \\ H_2(j\omega_k)C_2(j\omega_k) \\ H_3(j\omega_k)C_3(j\omega_k) \end{bmatrix}}_{\mathbf{m}}, \quad (5.6)$$

which can be solved for the unknown complex quantities $\alpha_0(j\omega_k), \alpha_1(j\omega_k)$ and $\beta_0(j\omega_k)$.

5.3.2 Determining a Desired Response

Suppose that in a control design problem, the design specifications are given in the frequency domain. For example, the specifications can include a desired gain margin, phase margin and bandwidth. Based on these specifications, one may consider a desired closed loop frequency response, denoted by $H^*(j\omega)$. Equation (5.5) can then be solved for the controller $C^*(j\omega)$ as

$$C^*(j\omega) = \frac{H^*(j\omega)\beta_0(j\omega) - \alpha_0(j\omega)}{\alpha_1(j\omega) - H^*(j\omega)}, \quad (5.7)$$

which guarantees that the desired closed loop frequency response, $H^*(j\omega)$, is attained. In the next subsection, we summarize the design steps and explain how the frequency response $C^*(j\omega)$ can be used to solve a controller design problem.

5.3.3 Steps to Controller Design

Here, we summarize the design steps toward solving a general control design problem based on the results provided in this chapter. Suppose that the frequency response data of plant P, denoted by $P(j\omega)$, is available, and the design objective is to find a controller which guarantees the stability and a set of frequency-domain specifications of the closed loop system, such as gain margin, phase margin and bandwidth. One may take the following steps to design such controller.

1. Connect three stabilizing controllers, C_1, C_2 and C_3, to the control system and measure the corresponding closed loop frequency responses, $H_1(j\omega)$, $H_2(j\omega)$ and $H_3(j\omega)$.
2. Solve (5.6), at a finite set of frequencies, for the unknown complex quantities, $\alpha_0(j\omega)$, $\alpha_1(j\omega)$ and $\beta_0(j\omega)$.
3. Define a desired closed loop frequency response, $H^*(j\omega)$, based on the frequency-domain design specifications.
4. Calculate the corresponding frequency response $C^*(j\omega)$ using (5.7).
5. Realize $C^*(j\omega)$ using system identification methods. An alternative approach is to consider a fixed-structure controller, such as a PID controller, and solve a least-square problem to determine the controller parameters in the stabilizing set. Denote the realized controller by $C_r(s)$.
6. Check $C_r(s)$ for stability. If $P(j\omega)$ and $C_r(j\omega)$ satisfy certain conditions at specific frequencies, then closed loop stability is guaranteed.
7. If $C_r(s)$ is not a stabilizing controller go to step 3, define a new $H^*(j\omega)$, and repeat steps 4 through 6 until the realized controller $C_r(s)$ is a stabilizing controller.

5.4 An Example of Control System Design

This example illustrates how the design method provided in this chapter can be used to find a controller, for an unknown linear plant, guaranteeing stability and performance of the closed loop system.

Problem. Suppose that the frequency response data of an unknown linear plant is as shown in Fig. 5.3, and the controller to be designed has a PID structure,

$$C(s) = \frac{k_d s^2 + k_p s + k_i}{s}. \tag{5.8}$$

Also, suppose that the required closed loop frequency-domain specifications are as follows:

1. Bandwidth ≈ 10 rad/sec,
2. PM > 100 deg.

The complete set of stabilizing PID controllers can be constructed as shown in Fig. 5.4 by using this frequency response.

Approach. The design objective is to find the "best" PID controller from the stabilizing set (Fig. 5.4) which guarantees the required specifications. Such controller can be designed following the steps given in the previous section:

1. Select 3 arbitrary controllers from the stabilizing set (Fig. 5.4), as follows:

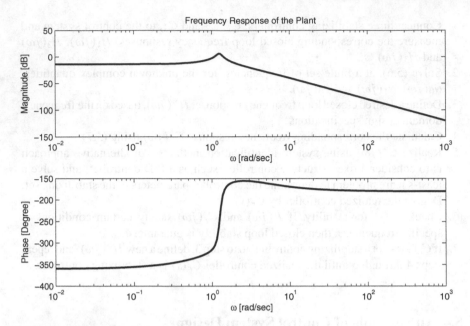

Fig. 5.3 Frequency response of an unknown linear plant

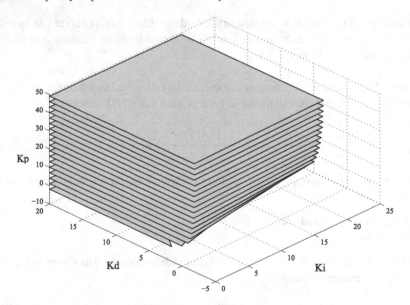

Fig. 5.4 The complete set of stabilizing PID controllers for this example

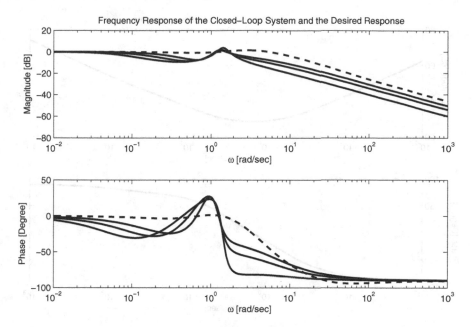

Fig. 5.5 Frequency response of the closed loop system $H_1(j\omega)$, $H_2(j\omega)$ and $H_3(j\omega)$ (*solid lines*) after embedding the controllers in (5.9) and the desired response $H^*(j\omega)$ (*dashed line*)

$$C_1(s) = \frac{s^2 + 2s + 0.5}{s},$$

$$C_2(s) = \frac{2s^2 + 2s + 1}{s},$$

$$C_3(s) = \frac{3s^2 + 2s + 2}{s}, \tag{5.9}$$

and place them in the closed loop system. For each case, measure the closed loop frequency response, $H(j\omega)$, as shown in Fig. 5.5 (solid lines).

2. Solve (5.6) for $\alpha_0(j\omega), \alpha_1(j\omega)$ and $\beta(j\omega)$, at a set of frequencies ω_k, $k = 1, 2, \ldots, N$.
3. Based on the provided design specifications, we define a desired closed loop frequency response, $H^*(j\omega)$, as plotted with the dashed line in Fig. 5.5.
4. $C^*(j\omega)$ is calculated using (5.7) as depicted in Fig. 5.6.
5. The frequency response $C^*(j\omega)$ is realized by a PID controller transfer function using MATLAB System Identification Toolbox, as:

$$C_r(s) = \frac{5s^2 + 39.8s + 13.3}{s}. \tag{5.10}$$

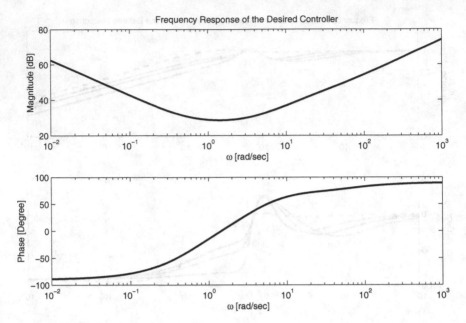

Fig. 5.6 Frequency response $C^*(j\omega)$

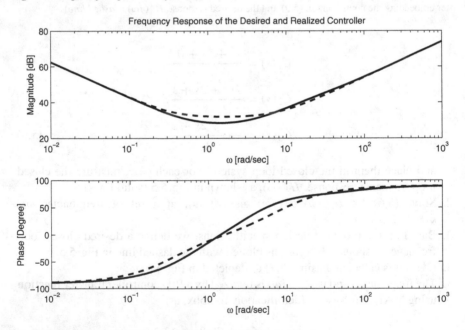

Fig. 5.7 Realization of $C^*(j\omega)$ (*solid line*) by $C_r(s)$ (*dashed line*)

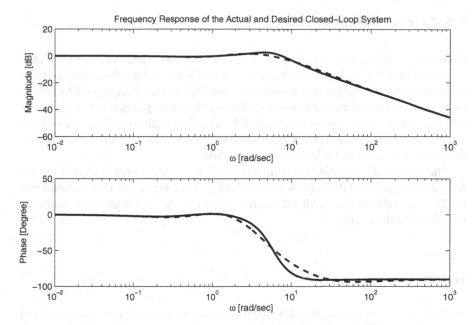

Fig. 5.8 The desired frequency response $H^*(j\omega)$ (*dashed line*) and the closed loop response after connecting $C_r(s)$ (*solid line*)

Fig. 5.7 shows the frequency response of $C^*(j\omega)$ (solid line) and $C_r(s)$ (dashed line).

6. Referring to the complete stabilizing set (Fig. 5.4), it can be easily verified that $C_r(s)$ is a stabilizing controller; hence, it is a solution to our control design problem.

The PID controller obtained in (5.10) is embedded into the system and the closed loop frequency response is measured as shown (with solid line) in Fig. 5.8 (the dashed line represents $H^*(j\omega)$). It can be verified (from Fig. 5.8) that the closed loop system constructed by connecting $C_r(s)$ in (5.10) has the following bandwidth and phase margin:

1. Bandwidth = 10.4 rad/sec,
2. PM = 104 deg.

Therefore, the controller $C_r(s)$ in (5.10), designed through this new measurement based approach, guarantees the stability and the required design specifications of the closed loop system.

5.5 Notes and References

Controller design usually requires a model of the plant. Here we have presented some results where it is shown that a design problem can be solved directly from frequency response measurements without knowing a model. This approach consists of connecting several controllers and measuring closed loop frequency responses at a finite number of frequencies. The controller transfer function can then be extracted for a given set of performance specifications. These preliminary results need to be developed into solid and effective design method.

The method of computing the complete set of stabilizing PID controllers from the frequency response of the plant is provided in [1]. Also, see [2] for more information on Nyquist stability of a feedback control system using the frequency response of the plant and the controller.

References

1. Bhattacharyya SP, Datta A, Keel LH (2009) Linear control theory: structure, robustness, and optimization. CRC Press, Boca Raton
2. Keel LH, Bhattacharyya SP (2010) A bode plot characterization of all stabilizing controllers. IEEE Trans Autom Control 55(10):2650–2654

Appendix

A.1 Current Control Using Two Resistances

Recall the proof of Theorem 2.2. We considered two cases: (1) $i \neq j, k$ and (2) $i = j$ or $i = k$.

Case 1: $i \neq j, k$

Suppose that $|\mathbf{M}| = 0$ in (2.26),

$$
\begin{vmatrix}
1 & R_{j1} & R_{k1} & R_{j1}R_{k1} & -I_{i1} & -I_{i1}R_{j1} & -I_{i1}R_{k1} \\
1 & R_{j2} & R_{k2} & R_{j2}R_{k2} & -I_{i2} & -I_{i2}R_{j2} & -I_{i2}R_{k2} \\
1 & R_{j3} & R_{k3} & R_{j3}R_{k3} & -I_{i3} & -I_{i3}R_{j3} & -I_{i3}R_{k3} \\
1 & R_{j4} & R_{k4} & R_{j4}R_{k4} & -I_{i4} & -I_{i4}R_{j4} & -I_{i4}R_{k4} \\
1 & R_{j5} & R_{k5} & R_{j5}R_{k5} & -I_{i5} & -I_{i5}R_{j5} & -I_{i5}R_{k5} \\
1 & R_{j6} & R_{k6} & R_{j6}R_{k6} & -I_{i6} & -I_{i6}R_{j6} & -I_{i6}R_{k6} \\
1 & R_{j7} & R_{k7} & R_{j7}R_{k7} & -I_{i7} & -I_{i7}R_{j7} & -I_{i7}R_{k7}
\end{vmatrix} = 0, \qquad \text{(A.1)}
$$

then it can be concluded that $\tilde{\beta}_3 = 0$ in (2.24). In this case, $\tilde{\beta}_2 = 0$ in (2.24), if

$$
|\mathbf{M}'| =
\begin{vmatrix}
1 & R_{j1} & R_{k1} & R_{j1}R_{k1} & -I_{i1} & -I_{i1}R_{j1} \\
1 & R_{j2} & R_{k2} & R_{j2}R_{k2} & -I_{i2} & -I_{i2}R_{j2} \\
1 & R_{j3} & R_{k3} & R_{j3}R_{k3} & -I_{i3} & -I_{i3}R_{j3} \\
1 & R_{j4} & R_{k4} & R_{j4}R_{k4} & -I_{i4} & -I_{i4}R_{j4} \\
1 & R_{j5} & R_{k5} & R_{j5}R_{k5} & -I_{i5} & -I_{i5}R_{j5} \\
1 & R_{j6} & R_{k6} & R_{j6}R_{k6} & -I_{i6} & -I_{i6}R_{j6}
\end{vmatrix} = 0, \qquad \text{(A.2)}
$$

and $\tilde{\beta}_1 = 0$ in (2.24), if

S. P. Bhattacharyya et al., *Linear Systems*, SpringerBriefs in Applied Sciences and Technology, DOI: 10.1007/978-81-322-1641-4, © The Author(s) 2014

$$|\mathbf{M}''| = \begin{vmatrix} 1 & R_{j1} & R_{k1} & R_{j1}R_{k1} & -I_{i1} & -I_{i1}R_{k1} \\ 1 & R_{j2} & R_{k2} & R_{j2}R_{k2} & -I_{i2} & -I_{i2}R_{k2} \\ 1 & R_{j3} & R_{k3} & R_{j3}R_{k3} & -I_{i3} & -I_{i3}R_{k3} \\ 1 & R_{j4} & R_{k4} & R_{j4}R_{k4} & -I_{i4} & -I_{i4}R_{k4} \\ 1 & R_{j5} & R_{k5} & R_{j5}R_{k5} & -I_{i5} & -I_{i5}R_{k5} \\ 1 & R_{j6} & R_{k6} & R_{j6}R_{k6} & -I_{i6} & -I_{i6}R_{k6} \end{vmatrix} = 0. \tag{A.3}$$

Therefore, the results for this case can be summarized as follows:

- if $|\mathbf{M}| = 0$, $|\mathbf{M}'|$, $|\mathbf{M}''| \neq 0$:

$$I_i(R_j, R_k) = \frac{\alpha_0 + \alpha_1 R_j + \alpha_2 R_k + \alpha_3 R_j R_k}{\beta_0 + \beta_1 R_j + R_k}. \tag{A.4}$$

- if $|\mathbf{M}| = |\mathbf{M}'| = 0$, $|\mathbf{M}''| \neq 0$:

$$I_i(R_j, R_k) = \frac{\alpha_0 + \alpha_1 R_j + \alpha_2 R_k + \alpha_3 R_j R_k}{\beta_0 + R_j}. \tag{A.5}$$

- if $|\mathbf{M}| = |\mathbf{M}''| = 0$, $|\mathbf{M}'| \neq 0$:

$$I_i(R_j, R_k) = \frac{\alpha_0 + \alpha_1 R_j + \alpha_2 R_k + \alpha_3 R_j R_k}{\beta_0 + R_k}. \tag{A.6}$$

- if $|\mathbf{M}| = |\mathbf{M}'| = |\mathbf{M}''| = 0$:

$$I_i(R_j, R_k) = \alpha_0 + \alpha_1 R_j + \alpha_2 R_k + \alpha_3 R_j R_k. \tag{A.7}$$

For each case above the constants can be determined using measurements.

Case 2: $i = j$ or $i = k$

If $|\mathbf{M}| = 0$ in (2.29),

$$|\mathbf{M}| = \begin{vmatrix} 1 & R_{k1} & -I_{i1} & -I_{i1}R_{j1} & -I_{i1}R_{k1} \\ 1 & R_{k2} & -I_{i2} & -I_{i2}R_{j2} & -I_{i2}R_{k2} \\ 1 & R_{k3} & -I_{i3} & -I_{i3}R_{j3} & -I_{i3}R_{k3} \\ 1 & R_{k4} & -I_{i4} & -I_{i4}R_{j4} & -I_{i4}R_{k4} \\ 1 & R_{k5} & -I_{i5} & -I_{i5}R_{j5} & -I_{i5}R_{k5} \end{vmatrix} = 0, \tag{A.8}$$

then $\tilde{\beta}_3 = 0$ in (2.27). The following cases are possible: $\tilde{\beta}_2 = 0$ in (2.27), which happens if

$$|\mathbf{M}'| = \begin{vmatrix} 1 & R_{k1} & -I_{i1} & -I_{i1}R_{j1} \\ 1 & R_{k2} & -I_{i2} & -I_{i2}R_{j2} \\ 1 & R_{k3} & -I_{i3} & -I_{i3}R_{j3} \\ 1 & R_{k4} & -I_{i4} & -I_{i4}R_{j4} \end{vmatrix} = 0, \tag{A.9}$$

and $\tilde{\beta}_1 = 0$ in (2.27), if

$$|\mathbf{M''}| = \begin{vmatrix} 1 & R_{k1} & -I_{i1} & -I_{i1}R_{k1} \\ 1 & R_{k2} & -I_{i2} & -I_{i2}R_{k2} \\ 1 & R_{k3} & -I_{i3} & -I_{i3}R_{k3} \\ 1 & R_{k4} & -I_{i4} & -I_{i4}R_{k4} \end{vmatrix} = 0. \qquad (A.10)$$

For this case we can state the results as follows:

• if $|\mathbf{M}| = 0$, $|\mathbf{M'}|$, $|\mathbf{M''}| \neq 0$:

$$I_i(R_j, R_k) = \frac{\alpha_0 + \alpha_1 R_k}{\beta_0 + \beta_1 R_j + R_k}. \qquad (A.11)$$

• if $|\mathbf{M}| = |\mathbf{M'}| = 0$, $|\mathbf{M''}| \neq 0$:

$$I_i(R_j, R_k) = \frac{\alpha_0 + \alpha_1 R_k}{\beta_0 + R_j}. \qquad (A.12)$$

• if $|\mathbf{M}| = |\mathbf{M''}| = 0$, $|\mathbf{M'}| \neq 0$:

$$I_i(R_j, R_k) = \frac{\alpha_0 + \alpha_1 R_k}{\beta_0 + R_k}. \qquad (A.13)$$

• if $|\mathbf{M}| = |\mathbf{M'}| = |\mathbf{M''}| = 0$:

$$I_i(R_j, R_k) = \alpha_0 + \alpha_1 R_k. \qquad (A.14)$$

A.2 Current Control Using Gyrator Resistance

Recalling the proof of Theorem 2.4, we considered two cases: (1) The i-th branch is not connected to either port of the gyrator, and (2) The i-th branch is connected to one port of the gyrator.

Case 1: The i-th branch is not connected to either port of the gyrator
Here, we consider the case where $|\mathbf{M}| = 0$ in (2.36),

$$|\mathbf{M}| = \begin{vmatrix} 1 & R_{g1} & R_{g1}^2 & -I_{i1} & -I_{i1}R_{g1} \\ 1 & R_{g2} & R_{g2}^2 & -I_{i2} & -I_{i2}R_{g2} \\ 1 & R_{g3} & R_{g3}^2 & -I_{i3} & -I_{i3}R_{g3} \\ 1 & R_{g4} & R_{g4}^2 & -I_{i4} & -I_{i4}R_{g4} \\ 1 & R_{g5} & R_{g5}^2 & -I_{i5} & -I_{i5}R_{g5} \end{vmatrix} = 0. \qquad (A.15)$$

This implies that $\tilde{\beta}_2 = 0$ in (2.34). Also, $\tilde{\beta}_1 = 0$ in (2.34), if

$$|\mathbf{M}'| = \begin{vmatrix} 1 & R_{g1} & R_{g1}^2 & -I_{i1} \\ 1 & R_{g2} & R_{g2}^2 & -I_{i2} \\ 1 & R_{g3} & R_{g3}^2 & -I_{i3} \\ 1 & R_{g4} & R_{g4}^2 & -I_{i4} \end{vmatrix} = 0. \tag{A.16}$$

The results for this case can be summarized as:

- if $|\mathbf{M}| = 0$, $|\mathbf{M}'| \neq 0$:

$$I_i(R_g) = \frac{\alpha_0 + \alpha_1 R_g + \alpha_2 R_g^2}{\beta_0 + R_g}. \tag{A.17}$$

- if $|\mathbf{M}| = |\mathbf{M}'| = 0$:

$$I_i(R_g) = \alpha_0 + \alpha_1 R_g + \alpha_2 R_g^2. \tag{A.18}$$

Case 2: The i-th branch is connected to one port of the gyrator
Here, suppose that $|\mathbf{M}| = 0$ in (2.39),

$$|\mathbf{M}| = \begin{vmatrix} 1 & R_{g1} & -I_{i1} & -I_{i1}R_{g1} \\ 1 & R_{g2} & -I_{i2} & -I_{i2}R_{g2} \\ 1 & R_{g3} & -I_{i3} & -I_{i3}R_{g3} \\ 1 & R_{g4} & -I_{i4} & -I_{i4}R_{g4} \end{vmatrix} = 0. \tag{A.19}$$

Therefore $\tilde{\beta}_2 = 0$ in (2.37). In this case, $\tilde{\beta}_1 = 0$ in (2.37), if

$$|\mathbf{M}'| = \begin{vmatrix} 1 & R_{g1} & -I_{i1} \\ 1 & R_{g2} & -I_{i2} \\ 1 & R_{g3} & -I_{i3} \end{vmatrix} = 0. \tag{A.20}$$

We summarize the results for this case as follows:

- if $|\mathbf{M}| = 0$, $|\mathbf{M}'| \neq 0$:

$$I_i(R_g) = \frac{\alpha_0 + \alpha_1 R_g}{\beta_0 + R_g}. \tag{A.21}$$

- if $|\mathbf{M}| = |\mathbf{M}'| = 0$:

$$I_i(R_g) = \alpha_0 + \alpha_1 R_g. \tag{A.22}$$

About the Authors

Prof. S. P. Bhattacharyya is internationally renowned for his fundamental contributions to Control Theory. These include the solution of the multivariable servomechanism problem, robust and unknown input observers, pole assignment algorithms, robust stability under parametric uncertainty, fragility of high order controllers and Modern PID control. He has coauthored 7 books and over 200 papers in the Control field. His current research focus is on model free approaches to engineering design, the topic of this monograph. Bhattacharyya is the Robert M. Kennedy Professor of Electrical and Computer Engineering at Texas A&M University. He is an IEEE Fellow, an IFAC Fellow and a Foreign Member of the Brazilian Academy of Sciences. He received the B. Tech degree from IIT Bombay in 1967 and the Ph.D. degree from Rice University in 1971, under the supervision of the late Prof. J. B. Pearson.

Prof. L. H. Keel is a renowned expert in Control Systems research. His research interest includes linear systems theory, complexity Issues in controller design, and robust control. He has co-authored 3 books and approximately 200 journal and conference papers in these areas. He is currently a Professor of Electrical and Computer Engineering at Tennessee State University in Nashville, Tennessee, USA. He received his B.S. from Korea University in 1978, and Ph.D. degree from Texas A&M University in 1986.

D. N. Mohsenizadeh is a Ph.D. candidate in the Department of Mechanical Engineering at Texas A&M University. He received his B.S. degree in mechanical engineering from the Shiraz University, Iran in 2007 and his M.S. degree in mechanical engineering from the Texas A&M University in 2010. He has 2 peer-reviewed publications in highly respected scientific journals as well as 8 conference papers. His research interests are in measurement based approaches to linear systems, synthesis of fixed structure controllers and linear control theory. His interests also lie in the field of structural analysis, where he has worked with the Boeing Space & Intelligence Systems on designing a new generation of solar arrays.

S. P. Bhattacharyya et al., *Linear Systems*, SpringerBriefs in Applied Sciences and Technology, DOI: 10.1007/978-81-322-1641-4, © The Author(s) 2014

About the Book

This brief presents recent results obtained on the analysis, synthesis and design of systems described by linear equations. It is well known that linear equations arise in most branches of science and engineering as well as social, biological and economic systems. The novelty of this approach is that no models of the system are assumed to be available, nor are they required. Instead, a few measurements made on the system can be processed strategically to directly extract design values that meet specifications without constructing a model of the system, implicitly or explicitly. These new concepts are illustrated by applying them to linear DC and AC circuits, mechanical, civil and hydraulic systems, signal flow block diagrams and control systems. These applications are preliminary and suggest many open problems. The results presented in this brief are the latest effort in this direction and the authors hope these will lead to attractive alternatives to model-based design of engineering and other systems.

S. P. Bhattacharyya et al., *Linear Systems*, SpringerBriefs in Applied Sciences and Technology, DOI: 10.1007/978-81-322-1641-4, © The Author(s) 2014

Index

A
A Measurement Theorem, 15

B
Bandwidth, 74
Block diagram, 72

C
Cramer's rule, 1
Current control, 18
Current phasor, 43

F
Fixed-structure controller, 75

G
Gain margin, 74
Generalization of Thevenin's Theorem, 23
Generalized Superposition Theorem, 12
Global stiffness matrix, 56
Gyrator, 29

H
Hydraulic network, 58

I
Impedance, 43
Interval design problem, 24

M
Mass-spring system, 53

Mechanical systems, 53
Monotonic behavior, 22

P
Parameterized solution, 11
Phase margin, 74
PID controller, 75
Pipe resistance, 58
Polynomial model, 5
Power level, 32
Power phasor, 43

R
Rank dependency, 3, 4
Rational model, 8

S
SISO control system, 73
Superposition Principle, 32
Superposition Theorem, 12

T
Thevenin's Theorem, 15, 23
Transfer function, 72
Truss structure, 55

U
Unknown linear AC circuit, 44
Unknown linear DC circuit, 20

V
Voltage phasor, 43

S. P. Bhattacharyya et al., *Linear Systems*, SpringerBriefs in Applied Sciences
and Technology, DOI: 10.1007/978-81-322-1641-4, © The Author(s) 2014